MOTIVES FOR METAPHOR IN SCIENTIFIC AND TECHNICAL COMMUNICATION

Timothy D. Giles
Georgia Southern University

Baywood's Technical Communications Series
Series Editor: CHARLES H. SIDES

Routledge
Taylor & Francis Group
LONDON AND NEW YORK

First published 2007 by Baywood Publishing Company, Inc.

2 Park Square, Milton Park, Abingdon, Oxon OX14 4RN
711 Third Avenue, New York, NY 10017, USA

Routledge is an imprint of the Taylor & Francis Group, an informa business

First issued in paperback 2017

Library of Congress Catalog Number: 2007006412
ISBN 13: 978-0-89503-337-6 (hbk)

Library of Congress Cataloging-in-Publication Data

Giles, Timothy D., 1958-
  Motives for metaphor in scientific and technical communication / by Timothy D. Giles.
      p. cm. -- (Baywood's technical communications series)
  Includes bibliographical references and index.
  ISBN: 978-0-89503-337-6 (cloth : alk. paper) 1. Communication of technical information.
2. Technical writing. 3. Metaphor. I. Title.

T11.G52 2007
601. '4--dc22

                                                            2007006412

ISBN 978-0-89503-337-6 (hbk)
ISBN 978-0-415-43440-9 (pbk)

For my parents,
David and Frances Giles

Desiring the exhilarations of changes:
The motive for metaphor, shrinking from
The weight of primary noon,
The A B C of being . . .

(Wallace Stevens, "The Motive for Metaphor")

Table of Contents

Acknowledgments

This book reflects an interest in metaphor in scientific and technical communication that I have pursued for over 20 years. I owe a number of individuals and institutions my thanks for their input and support during this project. Charles Sides, Baywood's Technical Communications Series editor was instrumental in this project arriving at fruition. Also at Baywood Publishing, I would like to thank the editorial staff, and especially Bobbi Olszewski and Julie Krempa, for their fine attention to detail.

Other institutions and individuals provided time and space to facilitate this project and allow the research and writing to take place. At the University of Minnesota, Victoria Mikelonis pointed me in the direction of much of the rhetorical theory that directs my analysis. Laura Gurak, Lee-Ann Kastman Breuch, and Sally Kohlstedt at the University of Minnesota read early drafts and made valuable comments. At Georgia Southern University, Mark Richardson also commented on early drafts and read page proofs. At East Carolina University, I recognize Keats Sparrow and Bertie Fearing for sewing my nascent interest in metaphor as it manifests itself in technical communication. At the Research Triangle Institute, I thank Joe Alexander for his many years of encouragement of my interests in science in general and in scientific and technical communication in particular.

Finally, let me thank my wife Nina Freifeld Giles and our daughter Anna. In *Classical Rhetoric for the Modern Student*, Edward P. J. Corbett dedicates the book to his wife and children, "whose persistent image of me is that of a man hunched over a typewriter." I hope that is not Nina and Anna's image of me; I haven't used a typewriter for years.

The Problem of Metaphor in Scientific and Technical Communication

A presentation at a recent meeting of the American Association for the Advancement of Science attempted to account for the factors affecting public perception of science, drawing no more tangible conclusions than that people are more likely to support scientific research if they know more about it. Even such a conclusion, however, cannot be directly credited to better science education (Pearson, 2005). The debate over cloning is an excellent example. After several years of debate, a United Nations legal committee has recommended that member nations ban all forms of human cloning, which includes research on cloning that could someday generate organs for those who need them to continue living. Some have suggested that scientists exploring cloning are playing God. If cloning is hampered and lives are lost as a result, then who, indeed, is playing God, those who clone, or those who prevent cloning? An examination of cloning reveals that no central metaphor has emerged that communicates cloning to the public. This book broadly recommends metaphors and analogies as an epistemological strategy that can be generative for the scientist, the engineer, and the lay audience, and specifically for metaphor and analogy to be taught in the scientific and technical communication classroom.

Metaphor and analogy have long aided and directed scientific thinking. In the seventeenth century Rene Descartes theorized that light was contained in a medium, which led to the theory of light as a wave. Newton's experiments with the prism later suggested that light was a particle, because the prism broke light into bands of color. Today, theory of light plays a role in developing technology, from eyeglasses to CRT screens to fiber optic cables. Metaphorically, industry thinks of light as a wave and as a particle, depending upon the application. In a fiber optic cable, industry thinks of light as a wave. For images to appear on

1

computer screens, industry must think of light as a particle. So for an image to appear on the computer screen via the Internet, a wave of data encased in light must travel down the fiber optic cable to the computer, where the signal is translated into particles that appear on the screen. In either case, it was necessary for light's wave-like properties or its particle-like properties to be recognized in order for theory to be shaped into usable products. Theories of light from the seventeenth century until now have proceeded and advanced by means of metaphor, evidence that metaphor has long been important to scientific writing, which this book reads, and defends, as a type of technical communication.

Does the shift between metaphors of light mean that there is a type of negative tension occurring here? To the contrary, such tension is not at all negative. Rather, it is a sign of health and can be read as science's growing pains. The title of Thomas Kuhn's *The Essential Tension* (1977) suggests instead that such stress is necessary, especially for a paradigm shift to occur. A good example of yet another paradigm shift for theories of light happened in the early nineteenth century when Thomas Young performed the double slit experiment. Young allowed a beam of light to pass through a hole in a screen. On a surface where the light struck, he made two holes, and on a surface behind that, he expected to see points of light where it had passed through the two holes. Instead, he saw bands of light, which contradicted what he expected, at least according to Newton's theory. Hence, Young reverted to Descartes' wave metaphor. Einstein's theory of relativity allows for the comprehension of light's wave-like and particle-like properties.

The passing and shifting of metaphors over centuries and from one scientist to the other further calls to mind Kuhn's influential work in science studies. He has argued that science is a social construction, and he has also carefully delineated the idea of paradigm shifts in science. Scientific metaphors are certainly further evidence of the social nature of science. As my study will indicate, it is difficult, if not impossible, to identify the coiner of the Solar System Analogy since, as I shall argue, it can be traced back to the ancient Greek atomists. As other theorists have noted, the value of studying metaphor in a scientific context is that metaphor is communal as it is passed from scientist to scientist, or from group of scientists to group of scientists, as the case study of the Solar System Analogy (SSA) evidences.

These examples demonstrate the value of metaphor to scientific thought, but what are the implications for technical communication, specifically in Information Technology (IT) and engineering? As a field, IT is rife with metaphors. E-mail and its accompanying desktop icons are obvious examples; and how is it that e-mail is sent, over the Internet, the Information Superhighway, or the World Wide Web? All of these are metaphors for tying down these abstract expressions of silicon and light.

Metaphor is important to engineering as well. A good example is John Smeaton's train of thought as he developed ideas for what became the Eddystone lighthouse. Prior to Smeaton's design of the Eddystone lighthouse, most lighthouses were built like Roman watchtowers, as wide at the bottom as at the top. Smeaton's journals reveal how he considered structuring the replacement lighthouse, the third one to stand on England's Eddystone reef. Two previous ones, the first built like a Roman watchtower and the second conical, had been swept away. Initially, Smeaton envisioned a lighthouse structured like a cradle so that it would rock with storms. Then he considered structuring it like a ship so that the lighthouse could ride the waves. However, it occurred to him that a cradle can tip over and a ship can capsize, so he settled on an oak tree structure, wider at the bottom than at the top, but tapering more gradually than a cone. The cradle and the ship were analogies that suggested the next steps, and some might read them as having been dispensed with at that point, except for how they led to the metaphor that inspired Smeaton to build a lighthouse that then stood on the England's Eddystone Reef for over two hundred years. Smeaton's lighthouse finally had to be moved inland because the rock around it had eroded to the point that it had become dangerous to use. This lighthouse, which still stands today, is a testament to the concrete value of metaphor as part of the engineer's thinking process (Smeaton, 1953, pp. 90-100).

Of course, metaphor can be problematic in science when metaphor becomes myth, which can occur when a scientist has too much invested in a metaphor and resists a paradigm shift. Einstein never fully accepted quantum mechanics because the idea of probability contradicted the precision of classical physics, and as a result, he was regarded as somewhat anachronistic in his later years (Greene, 2003). Other nineteenth-century physicists discussed in this book such as J. J. Thomson, who is credited with discovering the electron, and Oliver Lodge, who did important early work in radio and whose work is presented as an example of early twentieth-century science writing, would not relinquish the aether, the substance thought to pervade outer space, much like a thin atmosphere. Just as the metaphors for light shifted between Descartes, Newton, and Young to the current understanding, Einstein, Thomson, and Lodge may not have been completely wrong with their adherence to classical physics. Today, string theory seeks to combine the atomic with the cosmological realms in a spirit of unity that hearkens to Scottish Natural Philosophy, which this study will also explore for its connection with metaphoric thought in science. Dark matter in outer space may also renew the aether metaphor.

With the importance of metaphor to science and engineering established, some definitions are in order. I begin with that distinction between scientific and technical communication and then move to metaphor and analogy.

DIFFERENTIATING BETWEEN SCIENTIFIC AND TECHNICAL COMMUNICATION

It is helpful to attempt to differentiate between scientific writing and technical communication since they are so often mentioned in the same breath. How might they be defined? W. E. Britton (1965) surveyed a number of other scholars who had attempted to define technical writing. Throughout his article "What is Technical Writing?" he uses the terms "technical writing" and "scientific writing" synonymously and often conjunctively referring to them as "technical and scientific writing" (p. 114), something that he does 12 times in this article. He notes that others such as Blickle and Passe (1963) have defined technical writing as "writing that deals with subject matter in science, engineering, and business" (Britton, 1965, p. 113). Another approach is from the linguistic perspective, in terms of syntax and vocabulary, which Robert Hayes (1961) has defined inductively to the extreme. According to Britton, A. J. Kirkman's approach differentiates between technical writing and creative writing by naming the writing belonging to the fine arts as "associative writing," while technical writing is "sequential writing" (p. 114). Britton himself defines technical writing by its transparency. As an analogy, he compares aesthetic writing to a symphony. However, "technical and scientific writing can be likened to a bugle call," which illustrates, according to Britton, the idea that technical writing should have one meaning and one meaning only. He concludes by recommending that those who teach writing to science students should encourage them to write about the work in their discipline because "such an assignment not only is a real exercise in composition but also taxes the imagination of the student in devising illuminating analogies for effective communication" (p. 116). Such advice has not been fully realized, unfortunately.

Britton's definition is an early one, published in 1965. In general, but especially on the topic of metaphor in technical communication, I do not agree with Britton's assertion that technical writing should be defined in terms of its transparency, especially when it is used for epistemological purposes. For a metaphor to be generative, and by generative, I am thinking of McMullan's (1976) idea of fertility, it must allow the scientist to develop the metaphor in conjunction with, or as, a model. It is interesting, though, that Britton recommends the writing of analogies, which supports the basic idea of my work. His support of analogy undermines his assertion of technical writing as transparent, because for a metaphor to be used as I have described, it must be used consciously, and the more consciously it is used, the better, since scientific theory is typically generated through careful thought and study. Smeaton's metaphors were far from transparent. Instead, his journals record him rejecting the ship and cradle metaphors before arriving at the oak tree metaphor that allowed him to build the best lighthouse. Granted, some metaphors, such as ones related to IT, are most valuable for their transparency, but such metaphors are used for communication and to allow a lay audience to use computers, not to generate scientific theory.

As technical communication textbooks evidence (and I examine them in the next chapter), some technical communication scholars argue against the use of metaphor in general because of how it can be misinterpreted. These scholars are clearly still supporting the idea of technical communication as a transparent medium, a concept that Britton's early definition does not contain.

Not all technical communication scholars would agree with Britton (1965) on the issue of transparency. For example, Carolyn Miller (1979) has questioned in general the extent that technical writing can be considered transparent. To describe the argument opposing hers, she poses the windowpane metaphor to illustrate how many scientists view writing as something that is most valuable when it is transparent. She posits that to accept technical writing as transparent is to accept the positivist tradition apparent since the Seventeenth-Century Enlightenment that pigeonholes technical communication as a discipline without a subject, an idea that harkens to Socrates' admonitions against the sophists apparent especially in Plato's *Gorgias* (1990). More recently, Miller has noted, this tradition's position may be described as reinforcing the idea that "if language is highly decorative or opaque, then we see what is really not there or we see it with difficulty" (p. 612). The idea of language that is "highly decorative" as problematic in terms of how it may stand between the reader and knowledge is an issue that this book addresses.

Though these aspects of technical communication are important, they still do not define technical or scientific writing. David Dobrin (1983) has offered a definition of technical writing as "writing that accommodates technology to the reader" (p. 242). He defines scientific writing as writing that makes truth claims that are responsive to the scientific discourse community, and he differentiates between scientific and technical writing in that technical writing can make truth claims relative only to a specific context. As an example, he poses, "'Nut A fits on bolt B,' does not refer to all the rest of the discourse. If the statement were found to be ineffective rather than invalid (but how would one invalidate it?), the rest of the discourse would still stand" (p. 231). For this reason, Dobrin contends that any connection between scientific writing and technical writing is weak. After differentiating between the two, he does not further pursue scientific writing, other than to note that, "In the scientific community, it would be considered an evasion of responsibility for a scientist to leave his or her writing to a scientific writer. (The only professional writing having to do with science . . . is science writing, a species of journalism)" (pp. 243-244). Given Dobrin's criticism of Britton and others who wrote technical writing definitions that make sweeping generalizations, it seems odd that he would ignore the many books and articles written by scientists and science writers each year that are intended for a general audience. However, his intent is to write a definition of technical writing, not scientific writing, so he does well to limit himself.

Both technical communication and scientific writing share a common goal to communicate to a specified audience. Though a technical communicator is more likely to write to appeal to a general audience, scientific writing can be aimed at a variety of audiences. It may have as its audience other scientists with highly specialized knowledge that allows vocabulary to create shortcuts that truly do communicate more effectively to an audience with a relevant background, but will be less meaningful, and often meaningless, to the general audience. The authors of these types of communications, which are most frequently journal articles, are usually scientists reporting on original research or raising questions through articles reviewing the work of their peers. However, science writing can also take the shape of articles designed to communicate with scientists in other fields but who do not have specialized knowledge. A molecular biologist may indeed be interested in a physicist's research, and of course, there are science journalists who specialize in communicating the discoveries of science to the general audience. The two case studies in this book deal with writing by scientists and science journalists. In terms of the work written by scientists, the articles studied here can be categorized as those written for other scientists with highly specialized knowledge in molecular biology and atomic physics; articles written for other scientists who lack the knowledge of these specialties; and articles written for the general public. Therefore, because of the breadth of the audience approached by these types of science writing, it is appropriate to discuss science writing in conjunction with technical communication. Because the articles in the case studies touch such a variety of audiences, they are appropriate to consider as manifestations of technical communication.

METAPHOR AND ANALOGY

This book uses the terms "metaphor" and "analogy" interchangeably. In some ways it would be more accurate to refer to the focus as "tropes and figures," as Fahnstock (1999) has carefully delineated. However, the term "metaphor" is so commonly played fast and loose in metaphor studies that this book refers to usages such as metonymy, synecdoche, personification, and many others, as metaphors, though each type of trope is discussed and defined within the appropriate context. The point is not to differentiate between how each plays out in an A:B versus an A:B::C:D structure, but to study how metaphor consciously and unconsciously affects science; this study argues that metaphor is epistemologically generative, but that science too often fancies that it has laid metaphor by the wayside like so much ornamentation. Unfortunately, a metaphor used unconsciously can misdirect science. In cloning research, for example, E. F. Keller (2000) has proposed that the metaphor "reprogram" may have misdirected research, especially when "reprogram" is used as a verb with the cell's nucleus as its object rather than the genome, which is part of the chromatin, the material surrounding the nucleus and to which research has finally turned. Consciousness of metaphor usage can prevent such a problem. However, such a potential problem is not a mandate against metaphor. Just as a technical communicator would not write a set of chainsaw instructions without warnings, self-conscious use of metaphor can allow it to be a powerful tool.

SUMMARY OF CHAPTERS

This book presents the argument for teaching metaphor in the technical writing classroom because of the value that can be accorded to metaphor as an epistemologically generative tool. Scientists use metaphor quite freely, but largely unconsciously, as the case study of current cloning research indicates. As a result, they sometimes create problems for themselves when communicating with the public. More importantly, metaphor has epistemological significance, as the first case study of the role of the Solar System Analogy (SSA) demonstrates. In this case, the analogy drawn between the solar system and the structure of the atom was abandoned for reasons more cultural, I argue, than epistemological. If Niels Bohr (1913) and Ernest Rutherford (1911) had been more consciously aware of the role metaphor can play, then the SSA could have continued to play a role in the development of theories of atomic structure. Indeed, the SSA is still apparent in Bohr's metaphors after he largely dispensed with the analogy. The case studies in this book support teaching metaphor in the technical communication classroom because future scientists, engineers, and technical communicators could benefit from becoming aware of metaphor and learning how to use it consciously.

Chapter One introduces the topic by drawing on the historical and pedagogical. It examines technical communication textbooks to build a case for a disparity in what students may be currently taught about metaphor and its role in technical communication.

Chapter Two argues for teaching metaphor in the technical communication classroom. This chapter reviews the literature in technical communication from different technical communication theoretical perspectives.

The dialogue regarding metaphor in technical communication scholarship spans nearly 30 years. During that time, metaphor has been a humanities concern in technical communication, but for the technical communication classroom, it can be most closely related to the computer industry. Though J. S. Harris (1975, 1986, 1993) kept the discussion alive by publishing an article on metaphor in scientific and technical communication about every 10 years, his approach is largely inductive and touches only lightly on theory. An important point in this chapter is that the question of metaphor in scientific and technical communication persists. It could be reasonably questioned whether its discussion is only academic and a byproduct of scholarship with roots in the humanities, emanating particularly from those with degrees in English. However, the concept of metaphor is important to the computer industry in terms of saving time and money and communicating more clearly with customers, especially new ones, in addition to the way it contributes to scientific epistemology, the focus of this book's argument.

On the other hand, the range of instruction on metaphor provided by introductory technical communication texts varies widely. Why are there chapters on

"Definition" or "Description" but not on metaphor? Instruction in the use of metaphor would be valuable to students in life sciences, physical sciences, and computer science. My study creates reason to provide room for such a chapter.

Chapter Three reviews the literature on the changing conceptions of metaphor, which can be drawn from a number of disparate fields, including the philosophical, literary, and rhetorical. With those influences in mind, I begin with the substitutionists and proceed to interaction theory and then to metaphor as an epistemology. With such theory as background, I then explore the case studies in the next two chapters. The focus of this chapter, regardless of the field that the works it reviews are drawn from, is on the rhetorical, the most advantageous perspective because the roots of metaphor as a theoretical construction lie in rhetoric.

Because Aristotle (1991) was the first to place a theoretical emphasis on metaphor, the discussion must begin with him. Though later classical scholars misinterpreted his theory of metaphor as advocating a substitutionist perspective, his work was influential and is evidenced in the writings of contemporary scholars. When Aristotle's theory of metaphor is examined carefully, it is more relevant to contemporary concerns. Understanding its nuances points to its universality as a concern, one that has perplexed people for ages.

Aristotle philosophized metaphor, though he retained it as part of his rhetoric, but not of his science. When Newton sanctified probability as a way of creating scientific knowledge, rhetoric was elevated, along with metaphor. However, for about 2,000 years after Aristotle, metaphor was taught as ornament. Evidence of it is in the work of the anonymous author of the *Rhetorica ad Herennium* (1990), and further evidence of the substitutionist approach is in I. A. Richards' (1936) work, though he set examination of metaphor's interaction as a goal. Chaim Perelman and Lucie Olbrechts-Tyteca (1969) begin to examine the way metaphor works, but they acquiesce the final word to a language of science, and they fail to acknowledge mathematics as another metaphor. Conversely, Nietzsche (1989, 1990) recognizes all language as metaphoric, and his influence can certainly be noted in Richards (1936) and Weaver (1990) as well as Perelman and Olbrechts-Tyteca, who resist Nietzsche. Paul Ricoeur (1975) corrects our notion of metaphor as a noun (an object) rather than as the verb (an interaction) that Aristotle intended, which sets the stage for the interactionists.

Before the interactionists are addressed, the tensionists are considered as a bridge. Their approach is perhaps best realized in the work of M. C. Beardsley and D. Berggren. Beardsley (1962) names the metaphorical moment in his identification of the metaphor's "twist." Berggren (1962/1963) is interested in this moment as well, but he disparages Beardsley's call for case studies, despite his lack of direction for research. The tensionists' work is but a prelude to that of the more fully realized interactionists.

Max Black (1962) is credited, and rightly so, with bringing the interactionist approach to the study of metaphor. Kuhn (1970) further contributes to metaphor's veracity by including it as another facet of the social construction of science. Black's focus on the verb as fertile ground for metaphor and the sentence as an organic whole sets the stage for a contemporary discussion of metaphor's philosophical dimensions that are explored epistemologically.

For metaphor as an epistemological construction, Karl Popper (1972) asserts that when a science such as psychology cannot cast a hypothesis that can be later borne empirically, that science is a pseudoscience. Its continued existence and appeal, then, become rhetorical. However, Popper's claim of such a science as psychology is not meant to be a searing indictment of its efficacy. Such a rhetorical stance can be epistemological since Popper recognizes the pseudoscience's value and contribution to society. On the other hand, this type of science cannot participate in verification through falsification, and the danger is if a pseudoscience becomes a dogma, or what Mary Hesse (1970) and others would call a myth. Metaphor, however, is part of the comparative nature of human thought. So long as it does not become dogmatic, it can be valuable as an epistemological tool.

Arbib and Hesse's (1986) approach to the epistemology of metaphor represents an important aspect of the current state of metaphor studies. Their research into artificial intelligence focuses on the layering of metaphors whose interaction creates a scenario where background knowledge can interact with metaphors in a structure with epistemological potential.

In addition, Hesse (1970) has noted the value of the role of historical research as it relates to a philosophy of metaphor, as has McMullan (1968, 1976), who also has differentiated between the *U*-fertility (unknown-fertility) and *P*-fertility (proven-fertility) of metaphor. All of these voices enrich the study of metaphor as a rhetorical act with epistemological significance. With epistemology, metaphor reaches the climax of its development in terms of its importance to science and philosophy.

Inconsistencies found in these voices lead to the impetus for the examination of the case studies in the next two chapters. Two questions emerge. One concerns whether or not mathematics is metaphorical. As an invented language, it would be interesting to know if it contains metaphors. Richards, along with Perleman and Olbrechts-Tyteca, posit that mathematics is not metaphorical. Deciding upon this issue has bearing upon the next question, which concerns whether or not all language is metaphorical, for Richards (1936), Black (1962), and Perelman and Olbrechts-Tyteca (1969) have declared all language to be metaphorical, but each insists upon breaking the metaphor down into two parts: a metaphorical part and a literal part. If there needs to be a literal part for the metaphor to be a metaphor, then there is a literal use of language that is not metaphoric. These questions are important because if all language is metaphorical, then the case for teaching metaphor in the technical communication classroom is strengthened.

Chapter Four is a case study that focuses on the analogy drawn between the solar system and the structure of an atom. Such a study is appropriate because it begins with a metaphor emerging into an analogy and ends in mathematics, as Black (1962) would have it, so it allows an exploration of whether or not mathematics is metaphorical. The study is drawn from the writings of three pairs of mid-nineteenth to early twentieth-century physicists as they attempt to determine the structure of the atom. The roots of metaphor in the work of these physicists are examined in light of Scottish Natural Philosophy. This case study illustrates how the metaphor serves a descriptive, explanatory, and predictive function that guides scientific theory and practice as well as serving as a teaching tool to disseminate scientific ideas to the public.

This particular slice of the history of science is important because this metaphor as it specifically relates to the structure of the atom has a definite beginning in the work of W. Thomson (Lord Kelvin) (1910) and J. C. Maxwell (1986) as the metaphor begins to take shape as the "vortex atom." Then it is followed in the work of J. J. Thomson (1907), who first proved the existence of the electron, and Oliver Lodge (1924), a physicist better known for his early work with electricity and radio but who extended the solar system metaphor to a concept he referred to as "atomic astronomy." Next, I examine the work of E. Rutherford (1911), who theorized that an atom contains a nucleus and quite a bit of empty space. Rutherford passed his work along to the young Niels Bohr (1913), who finally dispensed with the SSA when he felt the model could no longer contain ideas such as the leap of electrons from one orbit to another.

Examining the development of the SSA in the work of Kelvin (1910), Maxwell (1986), J. J. Thomson (1904), Lodge (1902), Rutherford (1911), and Bohr (1913) allows observation of the development of this analogy. Though it is more fre-quently referred to as the "Bohr Atom" or as the "Rutherford-Bohr Atom," but less often as the "Thomson Atom," Thomson worked with the SSA more frequently and over a longer period of time than either Rutherford or Bohr. Lodge contributed to its explication since it may be discerned in his work earlier than it appears in Thomson's. The SSA's nascence was ferreted out in the writings of Kelvin and Maxwell, whose educational experiences prior to Cambridge were influenced by Scottish Natural Philosophy. The Cambridge wranglers valued metaphor and analogy as well.

With Rutherford and Bohr, a cultural schism becomes apparent. The fact that they more readily rejected the SSA is indicative of a scientific cultural perspective that did not weigh metaphor with the same value accorded it in the British Isles. Certainly New Zealand was a British possession during Rutherford's time there, but it was far removed geographically from the environs of Cambridge, as well as the influences of Scottish Natural Philosophy. Bohr's intellectual influences regarded analogy as an aspect of a suspect materialism, and his dispensation of the SSA was regarded by many (Heilbron, 1985; Kuhn, 1993) as heralding a new science that dealt with quantitative expression without recognizing the quantitative as yet another metaphor, much less the more traditional application of metaphor. As a result, metaphor was swept aside as an anachronism.

Chapter Five examines the use of metaphor in articles published shortly after the announcement of the cloning of the sheep Dolly. Focusing on metaphors associated with cloning allows a consideration of whether or not all language is metaphorical. In this case, there is no central metaphor such as, "light is a wave," associated with cloning. However, a good metaphor would be helpful for communicating with the world outside of science, which is important in terms of securing funding as well as furthering theoretical and popular understanding of the phenomenon. Instead of conjuring the vision of the mad scientist in the lab, a good metaphor could create a context for cloning that could place it more effectively and less controversially in the public eye. There are, however, a number of tropes making the transition from metaphor to dead metaphor, which indicates the metaphoric nature of language.

An examination of secondary-school textbooks reveals no metaphor for cloning, unlike the examination of these texts related to physics that yielded not only the SSA as the most frequently used metaphor to describe the structure of the atom, but the most nearly accurate one. As a result, it may be concluded from its absence that there is no coherent, central metaphor for cloning, at least not one in popular use. If there were one, then, like the SSA, it would be reasonable to expect it to appear in a textbook.

With the cloning case study, a variety of tropes and figures are witnessed as scientists and science writers seek to express cloning's ramifications. Though no central metaphor emerges, technical metaphors affiliated with the computer industry are observed, and I am able to demonstrate other emerging metaphors that are transmogrifying into dead metaphors, which indicates the metaphoric nature of language.

I continue the examination of cloning by studying articles heralding the cloning of Prometea, the first cloned horse. What is most remarkable here is that she was cloned from a skin cell of the mare in whose womb she was then nurtured. These articles are examined for metaphoric usage to determine what changes might have occurred for the metaphors describing cloning in the six years after the publication of the articles on Dolly. The relevance of these studies is illustrated through the way that cloning is currently playing out in the international arena, where motions have appeared before the United Nations to either ban human reproductive cloning or to ban cloning altogether. The latter choice would inhibit research that would explore how to clone a single organ to replace a failing one.

In Chapter Six, I consider the implications of the theoretical literature, the two case studies, and the technical communication literature. I draw conclusions for the implementation of metaphor into the technical communication classroom.

This book offers evidence for why metaphor should be taught in the technical communication classroom as a rhetorical strategy. It contends that avoiding metaphor is a disservice to students who are preparing to become scientists, engineers, or technical communicators. Such scholarly exposition casts metaphor in both a contemporary and historical context that can strengthen the case for teaching metaphor in the scientific and technical communication classroom.

CHAPTER 1

Reintroducing Metaphor in the Technical Communication Classroom

A question I have posed to candidates seeking a technical and professional writing position concerns what they perceive as the difference between a business communication and technical communication course. Candidates who replied that they did not perceive much of a difference, other than the type of students, were bumped down in my estimation. To some degree, this book offers an answer to that question.

Though business, especially advertising, is rife with metaphors, they are largely rhetorical in the sense of being literally persuasive. The idea of e-mail, for example, is transparent and allows the business world to function without having to explain what is really going on when messages are created and coded.

For scientific thought, metaphors are epistemological as well as rhetorical. They are epistemological in the sense discussed in the Introduction for John Smeaton (1953) as he designed his lighthouse. An epistemological metaphor can also be rhetorical and can be instrumental in bringing about a Kuhnian paradigm shift, as the metaphors of light demonstrate.

The purpose of metaphor in business communication is to communicate and persuade. In scientific and technical communication, the purpose of metaphor is to communicate, but it is also a tool with generative epistemological significance, that, though perhaps not entirely absent from business communication, has been more readily evident in scientific and technical communication. The power of metaphor has not so much been explored for scientists as become a loose cannon. This study explores metaphor as a rhetorical strategy for scientific and technical communication.

PROBLEM

Metaphor has remained an active part of scientific and technical discovery, thought, and discourse whether it is oral or written. And well it should. Many philosophers and rhetoricians, including Friedrich Nietzsche (1990), I. A. Richards (1936), and Paul de Man (1979), have proclaimed all language as metaphoric. However, most science writing and technical communication textbooks have not assigned metaphor an important role. Typically, discussions of metaphor are relegated to chapters titled "Description" or "Definition." Why is there not a chapter entitled "Metaphor and Analogy?" Metaphor is important in scientific and technical communication, but its value is not readily apparent in texts. This book seeks to address this issue in four ways:

1. by reviewing the most recent developments in metaphor studies to establish a common understanding of how metaphors function (or malfunction) today;
2. by examining the use of the solar system model of atomic structure as an example of how a particularly fertile analogy guided both the theory and practice of science;
3. by conducting a rhetorical analysis of discussions of cloning, a contemporary scientific revolution that has been inhibited because it has not been framed by a central, coherent metaphor;
4. by demonstrating why metaphor should be taught in technical writing textbooks and how it can be used effectively to promote a greater understanding of scientific and technical concepts.

Furthermore, this book dispels the idea that metaphor in scientific and technical communication ended with Francis Bacon (Moran, 1985; Zappen, 1999), who has been interpreted as calling for a plain style in scientific and technical writing. Instead, this book examines the role that metaphor and analogy, as a part of Scottish Natural Philosophy, a seventeenth-century philosophical trend that extended itself well into the twentieth century, played in the development of theories of subatomic structure. Such an example provides the basis for teachers of technical communication to justify teaching metaphor and analogy in the technical communication classroom. To understand the context of metaphor in the teaching of technical communication, I examine how metaphor fares in technical communication textbooks and how early technical communication scholars thought about metaphor. Then I build the case for reintroducing metaphor.

METHODOLOGY

After reviewing the developments in metaphorical theory that provide the theoretical basis for the argument, I conduct a close rhetorical analysis of two case studies, comparing how the absence or presence of metaphor epistemologically shapes the studies. This book uses rhetorical analysis as part of the methodology for examining metaphor's epistemological value. Such rhetorical analysis includes examining various texts to identify metaphor and its use. Documents vary from scientific papers published in professional journals to articles intended for the general public to textbooks. Inconsistencies in the literature drawn from metaphor studies provide a focus for that examination. From what has been discerned about how metaphors operate in the case studies, I construct an argument for why metaphor should be reintroduced in the technical communication classroom.

SOME GENERAL CONSIDERATIONS OF METAPHOR

Prior to the twentieth century, a good metaphor or analogy was requisite to describing and generating theory. Scientific and, more recently, technical communication has long aspired to a plain style that is typically traced back to Sir Francis Bacon, who called for a direct correspondence between objects and ideas on the one hand and words on the other. As a result, it is assumed that if this plain style were not invoked in Bacon's lifetime, it was certainly in place by the time of the Royal Society, which was founded in 1660, 34 years after Bacon's death.

Other seventeenth-century voices derided the use of metaphor as well. Thomas Hobbes (1983) argued in 1651 against metaphors as illusions: "and reasoning based upon them, is wandering amongst innumerable absurdities; and their end, contention and sedition, or contempt" (pp. 116-117). As one of his "Absurd conclusions," Hobbes specifically lists metaphor seventh:

> the use of Metaphors, Tropes, and other Rhetoricall figures, in stead of words proper. For though it be lawful to say, (for example) in common speech, *the way goeth, or leadeth hither, or thither; the Proverb says this or that,* (whereas wayes cannot go, nor Proverbs speak;) yet in reckoning, and seeking of truth, such speeches are not to be admitted (pp. 114-115).

Other references to tropes can be found in his "Absurd conclusions." For example, the second one is, "the giving of names of *bodies,* to *accidents*; or of *accidents,* to *bodies*" (p. 114). As examples, he cites, "*Faith is infused,* or *inspired*; when nothing can be *powred,* or *breathed* into anything, but body; and that, *extension* is *body*; that *phantasmes* are *spirits,* &c" [Hobbes' italics], which suggests personification.

In 1690, J. Locke (1990) explained the untrustworthiness of metaphor, which he links specifically to rhetoric:

> But yet if we would speak of things as they are, we must allow that all the art of rhetoric, besides order and clearness; all the artificial and figurative

application of words eloquence hath invented, are for nothing but to insinuate the wrong ideas, move the passions, and thereby mislead the judgment; and so indeed are perfect cheats: therefore, however laudable or allowable oratory may render them in harangues and popular addresses, they are certainly, in all discourses that pretend to inform or instruct, wholly to be avoided; and where truth and knowledge are concerned, cannot but be thought a great fault, either of language or the person who uses them (p. 710).

Today, as a result, metaphor and analogy are not associated very closely with scientific and technical communication. Though most scientists are probably not well-versed in Bacon, Hobbes, or Locke, these philosophers created a tradition that was exacerbated by the day's educational experiences that rarely included anything resembling science and instead focused on learning Latin and Greek and memorization. A more immediate reason that can be easily remedied is the placement, absence, or warnings of metaphor's use in introductory technical communication textbooks, where budding scientists and engineers could be exposed to this process of thought and expression. Following the tradition Bacon established, many textbook writers consider metaphor to be an inexact form of expression.

TECHNICAL COMMUNICATION TEXTBOOKS

How metaphor is being taught in technical communication textbooks indicates how technical communicators are being introduced to metaphor as well as how it is being introduced to prospective scientists and engineers. Therefore, textbooks intended to introduce students to scientific and technical communication are most relevant.

Thirteen textbooks deemed appropriate for an introductory technical communication class were examined. They were qualified as appropriate for the introductory class through examination of their contents, which revealed a similarity in discussion of the writing process and the basics of technical communication, such as general instruction on letter and memo writing and specifics on typical technical communication rhetorical modes, such as instructions, process description, object description, and definition. The pattern of usage in these texts is displayed in Table 1. According to the table, the placement of the metaphor and analogy discussion differs from text to text, which indicates that as a field, technical communication has not decided where these rhetorical tools fit. A slight exception is made for L. J. Gurak's (2000) *Oral Presentations for Technical Communication* because, although it is focused on oral communication, it is a textbook soundly based, in general, in rhetorical theory, and it is the only text to include an entire chapter devoted to analogy and metaphor.

For the other texts, analogy and metaphor were most likely to be found in the "Definition" chapter or section of another chapter, and objections to analogy and

Table 1. Patterns of Placement of Discussion of
Metaphor and Analogy in Introductory
Technical Communication Texts

Chapter or section	Number of instances
Using Analogies to Explain Technical Ideas	1
Definition	7
International Audience	3
Description	2
Revision	2
Presenting Information	1
Visual Rhetoric	1
Figures of Speech	1
Logic	1
Organization	1
Persuasion	1
No Mention	1
Total:	22

metaphor were most likely to be found in a chapter or section on international communication. The fact that metaphor and analogy were mentioned in a variety of contexts indicates the topic's relevance. However, no textbook, other than the Gurak text, situated metaphor and analogy historically in scientific and technical thought, and none offered students any guidance on how to create them. In fact, there is very little differentiation between analogy and metaphor.

To further establish metaphor as a relevant topic for discussion, the extent of the discussion in the *Handbook of Technical Writing* (Alred, Brusaw, & Oliu, 2000) is notable. First, it is interesting that there is any discussion in the this text at all since, unlike the other textbooks cited here, the first sentence of its preface specifies that it is "a comprehensive resource for both academic and professional audiences" (p. vii). Organizationally, the *Handbook of Technical Writing* is encyclopedic in its alphabetic arrangement, and it has long been a primitive hypertext: throughout the book are boldfaced text that refer the reader to another entry. This book could just as easily be on a professional technical writer's desk as upon the student's, unlike the other books. Therefore, the diversity of metaphor and analogy discussion can be interpreted as either a professional concern or as an issue the authors wish to place before the professional's eye.

Some of these texts encourage the use of metaphor and analogy as a rhetorical strategy. The chapter in Gurak's (2000) *Oral Presentations for Technical Communication* is titled, "Using Analogy to Explain Technical Ideas" and is broken up into sections that define analogy, explain its power as a rhetorical tool, draw the connection with scientific and technical communication, and warn readers about potential pitfalls. To support the use of analogy, she cites numerous historical examples from Aristotle to Watson and Crick. Her suggestions for teaching analogy are referenced in Chapter Six of this book. Houp, Pearsall, Tebeaux, and Dragga's text (2002) mentions analogies in their chapter "Presenting Information," where they advise, "you should frequently use short, simple analogies, particularly when you are writing for lay people" (p. 179). They recommend analogy for extending a definition: "A voltmeter is an indicating instrument for measuring electrical potential. It may be compared to a pressure gauge used in a pipe to measure water pressure" (p. 186). Mike Markel (1998) provides two somewhat different examples of computer literacy: accessing a database and using a VCR or an ATM machine. The analogy becomes apparent when he compares the ability to drive an automobile with the ability to use a computer: computer users do not have to know RAM from ROM to be computer literate (p. 245). J. S. Lannon's (2003) discussion of analogy is brief, but he contributes to the discussion by noting the stylistic improvement an analogy can add to technical communication. His discussion falls in a chapter on revision. Analogies "sharpen the image," "emphasize a point," and "save words and convey vivid images." As an example, he notes that "The metal rod is inserted crosslike" can substitute for "The metal rod is inserted, **perpendicular to the long plane and parallel to the flat plane,** between the inner and outer sections of the clip" (p. 274). Lay, Wahlstrom, Selfe, Selzer, and Rude (2000) offer an extended analogy that compares the parts of a biological cell to a library. P. V. Anderson's (1999) treatment of analogy and metaphor is brief, though positive. As a definition of an analogy, he offers, "Example: An atom is like a miniature solar system in which the nucleus is the sun and the electrons are the planets that revolve around it" (p. 264), pointing to the ubiquity of the Solar System Analogy, which is discussed in greater detail in Chapter Four.

In all, nine texts recommended analogy and metaphor as rhetorical features valuable as communication tools. However, analogy and metaphor did not fare as well in the other four texts. Jones and Lane (2002) first mention analogy by defining it in a chapter titled "Achieving an Effective Style," but they warn writers about the danger of using metaphors because of the problems that metaphors may create upon translation. Later, analogy is mentioned in a section on extended definitions as one of nine ways of expanding upon a term, but without any direction on how to go about doing so. In an example of an extended definition of ionizing radiation, alpha particles are compared to "large and slow bowling balls," beta particles to "golf balls on a driving range," and gamma and x-radiation to "weightless bullets moving at the speed of light."

This selection in general, like much of physics, is rife with metaphors such as "a stream of electrons" (p. 514). K. R. Woolever (2002) mentions only metaphors, similes, and analogies under the heading, "International Style Guidelines," and, of course, recommends that writers use these approaches "with caution" (p. 137). Her technical description and process description sections are riddled with metaphor, such as this process description of a polariscope: "After leaving the first filter (the polarizer), the polarized light enters the transparent gear model and can only vibrate along two perpendicular planes coinciding with the planes of principal stress" (p. 220). "This variation in magnitude of principal stresses at points in the model causes the entire surface of the model to be covered with a number of 'fringes' (figure 4-2). From these fringes we can measure the maximum shear stress and stress direction" (p. 221). In their chapter "Audience Recognition and Involvement," S. J. Gerson and S. M. Gerson (2003) advise writers to "avoid figurative language," because it does not translate well. As an example of how this can be problematic, they offer the cliché, "The best offense is a good defense" (p. 75).

Finally, T. C. Kynell's (1999) case study approach to technical communication does not mention metaphor, analogy, or figurative language. However, in a section on "Technical Definition," an example of an extended definition of a fingerprint is rife with metaphors that are italicized, evidently to direct attention to the words "hooks," "forks," "eyes," and "islands," which describe the various shapes and patterns of the whorls (p. 101). Such use as an example without any direction on how to use metaphor (or even what one is) effectively is unfortunately typical rather than atypical.

There were also problems with accurate presentation of metaphor and analogy. The *Handbook of Technical Writing*'s (Alred, Brusaw, & Oliu, 2000) presentation was the least accurate. "False analogy" was defined as an instance of *post hoc, ergo propter hoc* (after this, therefore because of this), which is a logical fallacy and indicates confusion between cause and effect. A good example of *post hoc, ergo propter hoc* occurred in early 1980 when the economy surged shortly after newly-elected President Reagan took office, which led some to credit him with the upswing in the economy, when in fact he had not been in office long enough to have any effect on the economy; the improvement in the economy resulted from outgoing President Carter's actions. To credit President Reagan was an example of *post hoc, ergo propter hoc*, which is quite different from a false analogy. Any analogy, really, is a false one since an atom indeed is not a solar system, even when that analogy was thought to more accurately model the atom.

There were other problems with Alred, Brusaw, and Oliu (2000) as well. The metaphor "armlike" was called an analogy. Metaphors were classified as figures of speech, and in this section, the examples seemed culled from business exposition rather than scientific or technical writing.

Presentations of metaphor in other textbooks were problematic. W. S. Pfeiffer (2003) claimed three similes were analogies, labeling them as such in the margin.

Rather than analogies, the similes are "like a lawn mower," "like sawdust coming into contact with oil on a garage floor," and "like a vacuum cleaner." These similes are intended to demonstrate a company's process for cleaning up oil spills at sea (p. 147). Not only are they similes, but the imagery is somewhat mixed, which is not problematic in scientific and technical communication, but it does disallow calling it an analogy, especially when contrasted with the passage comparing a cell to a library cited in Lay et al. (2000). The only way each could be considered an analogy is in the sense that a simile is a truncated analogy, the same sense in which a metaphor is a truncated simile. To call it an analogy, these similes would have to be on the same topic. Such errors and the lack of direction, other than by exemplification, when it is accurate, point to the need for theoretical and practical background on this subject.

SCIENCE WRITING TEXTS

Often academic departments that teach technical communication courses also teach courses in science communication as well, so it is reasonable to delve into these types of textbooks to learn if metaphor is taught as a part of scientific writing.

The field of textbooks and other guides for science writing is much narrower than it is for technical communication. One of the best examples of a science writing text relevant to the field of technical communication is the second edition of A. M. Penrose and S. B. Katz's (2004) *Writing in the Sciences: Exploring Conventions of Scientific Discourse*. This book is part of the Allyn and Bacon technical communication series. Though analogy, metaphor, and simile are separate entries on the "List of Stylistic Features" page that follows the table of contents, their actual placement in the book suggests they are the rhetorical devices that dare not speak their name. Discussion of these rhetorical tools is buried in a chapter titled "Communicating with Public Audience," in section seven, "Adapting Through Comparison," which begins with remarking that comparison is helpful for creating effective explanations for a general audience. Then, as an illustration, Penrose and Katz introduce a definition of dinoflagellates as "twilight zone creatures: half-plant, half-animal . . . [that] move about, using their two flagella, or whiplike tails . . ." (p. 193). The next paragraph explains the value of synonyms before abruptly shifting to a discussion of simile and metaphor. Penrose and Katz illustrate their points about metaphor and simile by using other examples drawn from different instances that discuss dinoflagellates, and there is a reference to another part of the book that contains this passage in a set of articles on algae, but the authors do not point out the metaphoric use of "twilight zone creatures," which is definitely metaphoric because they use it to emphasize that the dinoflagellates are "half-plant, half-animal," not to describe any relationship dealing with time. The term "twilight zone creatures" seems to share more with popular culture than with science. On calling the tails "whiplike," the authors tell their readers, "This sort of definition in context is also known as *parenthetical explanation* . . . [or] apposition" (p. 193), but they do not pursue the metaphoric aspect.

While Penrose and Katz (2004) do not advise avoiding metaphor, they comment that it is stronger than simile and "a much more pervasive method of comparison than we might think" (p. 193). Ironically, they follow this statement with an example of the dinoflagellates as having a "Jekyll and Hyde personality," and note that a metaphor's meaning for an audience "is grounded in the audience's cultural knowledge" (p. 193), without commenting on the literary allusion of "Jekyll and Hyde personality."

They assert that, "Some researchers believe that metaphors actually structure, and to some extent, determine, the way we conceptualize the world" (p. 194), but their theoretical grounding for this statement is linguistic, rather than in rhetorical theory, as they cite only Lakoff and Johnson (1980) for theoretical background. Penrose and Katz (2004) note that, "it has been argued that metaphors are implicit in scientific models," and as evidence, they mention that "Niels Bohr's early models of the atom as a solar system immediately come to mind," which supports, again, the Solar System Analogy as one of my case studies. They conclude their discussion of metaphor by advising that metaphors should align with "the language, knowledge, and experience of your audience; do not merely switch to other technical metaphors embedded in your field that the audience will not understand," but they do not include a clear example of a technical metaphor. "Messenger RNA" could be a good example of technical metaphor meaningful to biologists that would not be useful to a general audience. They also advise that science writers should not mix metaphors, which is not supported with an example, nor is it consistent with the way the science is created epistemologically, where messenger RNA may be mentioned in the same breath as cumulus cells. On mixed metaphors, Lakoff and Johnson (1980) have commented,

> There is good reason why our conceptual systems have inconsistent metaphors for a single concept. The reason is that there is no one metaphor that will do. Each one gives a certain comprehension of one aspect of the concept and hides others. To operate only in terms of a consistent set of metaphors is to hide many aspects of reality (p. 221).

To resist mixed metaphors, according to Lakoff and Johnson, is to buy into the "objectivist" perspective that is positivistic and unrealistically accords to words a fixed definition.

Penrose and Katz conclude a brief discussion of analogy by telling their readers that "usually several strategies are used together to unpack a scientific concept; they may even be embedded in each other" (2004, p. 196), which suggests the epistemological value of metaphor, but with only an example that explains how chemicals react to one another by drawing an analogy between this process and how humans react to nourishment or companionship when they have lacked it. Explaining how physicists worked through the problem of the structure of the atom could better illustrate this point. The idea of an "orbit" was embedded in the Solar System Analogy, and it had to be dealt with for the theory to advance. Computer metaphors were embedded in cloning, and they may have been misdirecting research to the nucleus, rather than the chromatin. Penrose and Katz conclude with pointing their readers to several documents used as examples in the book to examine for evidence of metaphor.

Four other books on science writing were examined as well. Of the four, two did not mention metaphor or analogy (Friedland & Folt, 2000; Matthews, Bowen, & Matthews, 1997). The other two mentioned metaphor, but only briefly, and both discuss metaphor in chapters concerned with diction, which suggests a substitutionist approach. Of the two that discuss metaphor, the coverage of metaphor is split. Michael Alley (1996) posits that, "analogies are valuable . . . at conveying complex ideas or numbers." He follows his statement with a quote from Albert Einstein's *Relativity: The Special and General Theory,* which he illustrates with an analogy of a stone tossed from a moving train. Alley comments that "Einstein's analogy is so much more alive than the abstract question: Where do positions of an object lie in reality" (p. 116)? For this reason, Alley recommends analogies for making the quantitative more concrete. On the other hand, R. A. Day (1995) advises that although "we all love to use metaphors and other figures of speech . . . I urge you to use such devices sparingly . . . [because] whenever a word or phrase is used in other than its literal meaning, we risk losing the comprehension of our readers" (p. 22). He follows these admonitions with clichéd examples such as "roll over in his grave," which he describes as evidence of a writer's ability "to craft a beauty" of a metaphor (p. 23). He warns his audience against mixing metaphors, but another of his "beauties" is, "If this thing starts to snowball, it will catch fire everywhere" (p. 23). Mixed metaphors are not problematic for science writers, but the fact that Day warns his audience about them and then uses one as an example certainly mars his credibility.

These examples indicate that metaphor and analogy are not widely taught in science writing. As a matter of fact, a larger percentage of science writing guides did not mention metaphor at all. It is interesting that the most competent discussion occurred in a series of books devoted to technical communication. Still, even with the Penrose and Katz (2004) text, students are merely given a few examples. They are directed to engage in rhetorical analysis by identifying metaphoric usage, but they are not informed of what might be considered metaphorical. For example, there is a good deal of difference between how personification or analogy is recognized. These omissions provide further evidence for my argument.

To what extent can technical communication pedagogical theory account, or fail to account, for the current state of metaphor in technical communication textbooks? More recently, Johnson-Eilola, Selber, and Selfe (1999) have demarcated the development of technical communication pedagogy. In a sense, its further delineation may be found in J. A. Berlin's *Rhetoric and Reality: Writing Instruction in American Colleges 1900-1985* (1987) as he applies it to all writing instruction. Berlin interprets writing instruction in American colleges and universities as falling into the objective, subjective, or transactional camps with a focus on the chronological development of each.

With the objective, Berlin relates writing instruction at this stage with what he calls, "Scottish Common Sense Realism," which he argues influenced nineteenth-century American education as well. According to Johnson-Eilola, Selber, and Selfe (1999), the objective influence on technical communication is reflected in the ideals of correctness and transparency.

The subjective is characterized by social-construction approaches to writing manifested by peer group editing and group projects. At this point in the discussion, Johnson-Eilola, Selber, and Selfe (1999) comment on metaphor:

> Additionally, in technical communication classrooms, subjective theories of rhetoric support students' use of original and effective metaphorical language, which allows them to represent their rich consciousness and tacit understanding of the world in the most effective forms . . . . Thus, technical communication students are encouraged to help readers connect highly technical and unfamiliar processes with known and familiar processes (pp. 210-211).

The paragraph concludes with examples comparing scanning equipment to the human eye and traditional office equipment to computer data storage capabilities. The next paragraph discusses the implementation of computer technology into the writing process and concludes with "Finally, the subjective framework for understanding language and the world justifies the frequent use of metaphorical language" (p. 211).

These stages of development are presented as a prelude to the transactional, which recognizes the multiple experiences of readers and writers and multiple ways of making meaning. As a result, technical communication scholars are now concerned with issues such as "the study of ethics, responsible civic decision making, social policy formation, rhetorical theory, or risk communication" (Johnson-Eilola, Selber, & Selfe, 1999, p. 212). Metaphor would certainly seem to be a peer to any of these categories. The lack of consistency in the understanding of metaphor in technical communication texts may stem, indeed, from theory that considers metaphor as something that has been previously dealt with and neatly tucked away. Berlin, of course, carefully traces the roots of each of these trends in intellectual thought, recognizing, for example, that classical rhetoric is a part of transactional rhetoric (1987, p. 15). But the way that Johnson-Eilola, Selber, and Selfe have presented how Berlin's theory applies to technical communication suggests that metaphor is a concept that has been tacitly defined and contained. Ironically, they evidently do not seem to read metaphor as a part of rhetorical theory, which suggests they interpret metaphor from a classical perspective (post-Aristotelian, though) that regards metaphor as ornament rather than the epistemological act this study supports.

My research reveals the most opposition to metaphor in technical communication textbooks, where it appears in the name of international communication. Teaching students about metaphor should include warning them about the problems inherent in it as a rhetorical tool. Including information on falsification, for example, is a way in which metaphor may fail and yet yield the epistemologically fruitful. With international communication, so much more may interfere with communication than metaphor. Colors have cultural significance, for example. The sound of "nova" calls to the Spanish ear "No va" (it doesn't go), which heralded an unsuccessful car campaign in the Spanish-speaking world. For all languages, metaphors can be effective, but they must be carefully chosen, a fact to which the computer industry can attest, as the desktop metaphor I discuss in the next chapter indicates.

Textbooks evidence disagreement over the role of metaphor in technical communication. I have examined one pedagogical essay, but this matter certainly demands a closer examination of the technical communication literature to determine if it is typical or atypical.

CHAPTER 2

Metaphor in the Technical Communication Literature

This chapter examines technical communication theory to establish the relevance of this discussion of metaphor to the discipline. Broadly, this discussion can be categorized into technical communication, rhetoric, computer technology, the literary, historical, and pedagogical. Of course, technical communication draws from a variety of fields, so when I name a category "technical communication theory," I am thinking of theory that is developed specifically for our discipline. With pedagogical theory, for example, much of it concerns the teaching of writing, at least as it relates to this study, visual metaphor notwithstanding. Similar to the discussion of metaphor in technical communication textbooks, the theoretical literature is uneven in its presentation of this topic.

Discussion of metaphor among technical communication scholars indicates its relevance to the field. That there is a focus on metaphor in the computer industry provides an impetus for discussion in this chapter. Not all students in technical communication will be involved with the computer industry, but the chapters following this one will establish metaphor's potential relevance to the life sciences and physical sciences. As a result, teaching metaphor in the technical communication classroom has a much broader and stronger appeal.

While the general direction that discussion of metaphor has taken in scientific and technical communication is interesting, it is pertinent to consider a particular application, especially to the computer industry, since it has reinvented technical communication as a discipline.

METAPHOR AND THE COMPUTER

The computer industry is a technical communication arena where metaphor is important, for just as in physics, the manifestation of the abstractions of silicon pathways and particles of light must be named, and the act of naming, especially to describe the actions of computers, is metaphoric and extremely important for that reason. Visual icons in the computer industry have become important to the extent that they can be said to have a dollar value. As Gurak (2003) has pointed out, "Just as these language-based structures provide users with consistent informational cues, standardized icons will also provide users with clear maps of how to use information and products" (p. 493). On this note, Steve Bream offers an interesting metaphor to describe the usefulness of metaphor to the computer industry.

The title of Bream's "Metaphor Stacking and the Velveteen Rabbit Effect" (2000) contains his article's metaphoric cornerstones. To stack metaphors means to rely upon past metaphors to communicate with an audience. A good example is the desktop metaphor. When Macintosh shifted to OS/2, the desktop metaphor made the transition, and then it jumped ship to Windows. When it was appropriated for Windows, it no longer resembled a desktop. At that point, it became a velveteen rabbit, according to Bream, after the classic children's story of the same name. In "The Velveteen Rabbit," a toy rabbit becomes a real rabbit because it is loved, much in the same way that the desktop metaphor has traveled beyond the idea of a desktop. For developers, these metaphors are valuable because they communicate more effectively with an audience and therefore save time and money, but they are problematic for several reasons. First, to be effective, the new implementation must be exactly like the former one, or it becomes a problem because the new incarnation must be relearned each time it is used, rather than forcing the user to learn a new metaphor. Bream concludes that metaphor is a powerful but expensive tool.

Explaining how to write better computer documentation does not necessarily bequeath the writer credibility. Though R. M. Chisholm (1986) advises that metaphor should be used because of its value to the user, he suggests criteria for choosing metaphors useful to the document's audience to the extent that he thinks of metaphor as a way "to create a new language that will speak clearly to millions of new computer users" (p. 197). Such an approach is reminiscent of Bacon's goal to create a new language of correspondence that would be metaphoric in a substitutionist sense. Other rhetoricians such as I. A. Richards (1943) have attempted to create such a new version of English, only to meet with failure.

Chisholm discusses I. A. Richards' (1936) tenor and vehicle as Object X and Object Y. He notes Richards' idea of interaction as well as its application in Black's (1962) work, but his interest is not so much in how metaphor works as how it can be used effectively. To evaluate metaphor, he first suggests that technical communicators should consider whether a metaphor is necessary at all. If there is a good, one-to-one correspondence between an object and reality, then a metaphoric term may be unnecessary. Is "boot," Chisholm asks, a better term than "start," (1986, p. 207), especially when it can as easily mean to turn on a computer as to load a program?

Chisholm questions the extent to which a known word may be correctly understood in light of background experience. As an example, he cites the term "watershed," which designates an important turning point for historians, but could be misinterpreted as a covered bridge. Such a misinterpretation of a metaphor is part of the danger and supplies evidence for why metaphors should be carefully chosen.

Furthermore, Chisholm recommends that terms be metaphorically related. As an example, he notes that for word processing, the idea of "enter" is more similar to "return" since either can be on a keyboard and labeled as such. "Menu" and "scroll" are problematic, though, because neither is related to typing or data processing. It is worth noting that Chisholm made these observations in 1986. Terms such as "menu" and "scroll" are problematic because they "introduce a note of strangeness that makes the concept of word processing a bit less easy to grasp" (p. 210) and have since then transmogrified into dead metaphors. Aristotle (1991), on the other hand, has praised metaphor for introducing that note of strangeness, and for someone who is learning word processing, the words "scroll" and "menu," and what they mean to the new user, are far more familiar than the computer. Such an objection seems especially strange since Chisholm's article is rife with allusions to classical rhetoric, such as his reference to Plato's objection to "riddling metaphors and distorted meanings . . . [as] inappropriate for this kind of writing" (p. 211).

Chisholm recommends that the use of the metaphor be consistent. As examples, he notes that "type" and "enter" denote touching a key, but the functions can be quite different. He also asserts that the metaphor should be brief since any other use would belie it as a verbal shortcut. Conversely, the metaphor's acceptability is also important. As an example, he poses the term "abort" as problematic for the novice user whose connotations for a medical abortion might be negative.

Finally, Chisholm calls for the metaphor to be memorable. "Type" and "print" would be too ambiguous as metaphors, but again, these terms are more memorable for the word processor novice. Perhaps as a result, Chisholm does admit "menu" as a suitable metaphor, more concise than Table of Contents, which, interestingly enough, has become an online document's TOC.

Chisholm's discussion of metaphor for the computer industry suggests, at best, a substitutionist approach. He alludes to rhetorical theory, but it does not lend any type of unity to the discussion. His criteria for choosing metaphors are not useful from the perspective of providing insight into how and why metaphor works. Instead, he lists criteria for evaluation, and that evaluation is not terribly helpful. For example, the trashcan metaphor was problematic for MacIntosh as computers made the transition from the office to the home. Homeowners did not expect the computer to automatically empty the trash when the computer was shut down. This process had not been problematic in the office where employees are less likely to empty their own trashcans. While the trashcan at the office is emptied after hours by maintenance personnel, the trashcan at home is not emptied until someone who lives there takes it out. The problem it caused was that people who owned home computers sometimes lost files they would have liked to have had the option of later retrieving from the trash. Would Chisholm's criteria have identified the trashcan as problematic for home use? His criteria can be condensed to

1. necessity
2. comprehensiveness
3. metaphorical relationship
4. consistency
5. conciseness
6. neutrality
7. memorableness

First, is there a necessity for the term? Certainly there was some need for a metaphor to describe what should be done with items to be discarded.

Is the metaphor comprehensive? In this sense, Chisholm (1986) wondered to what extent a word can be understood against background experience. Certainly a trashcan would pose no problem.

Is there a reasonable metaphorical relationship between the words? Chisholm's concern here is that the metaphoric usage is not so far afield that it does not clearly explain that for which it is intended. As a place where material should be put that is no longer needed and should be purged from the system, a trashcan seems apt.

Is there a consistency of usage? The trashcan metaphor had been successfully used on office computers, and the transition was identical. All that changed was the shift from office environment to home environment. As a matter of fact, like the desktop metaphor, it can be read as a velveteen rabbit (Bream, 2000).

Is the usage concise? It certainly is.

Is it neutral? In terms of neutrality, Chisholm (1986) was thinking of positive or negative connotations associated with a term. Perhaps a trashcan carries some negative connotations, but it is an ordinary, ever-present item, one that does not carry the negative connotations of a somewhat similar item such as, say, a toilet.

Because a trashcan is such an omnipresent item, it can also be read as satisfying the final criterion, that of being memorable through its simplicity. According to Chisholm's criteria, then, a trashcan should have been a satisfactory metaphor. It created problems for home users, unfortunately, who represented a growing market, and as a result, the trash on the desktop and in e-mail programs must now be emptied manually.

Chisholm's (1986) criteria do not offer a useful evaluation. While they could be somewhat useful, they suggest that somehow they can contain and tame metaphors. Bream (2000) is correct when he assigns the characteristics of "powerful" and "expensive" to metaphors. The computer industry cannot function without them, which has been especially apparent since the early days of the Internet and e-mail. Determining which metaphors are the most appropriate is a complex task that includes research and testing. Coupled with a familiarity of the rhetorical theory of metaphor, such a program of study could provide a student with the necessary background to make intelligent decisions that save time and money.

Technical communication as a field is interested in metaphor from general humanistic and specific historical and rhetorical perspectives. Though metaphor's application can be useful to any science, exposing technical communication students specifically to metaphor as an aspect of the discipline's theory is especially useful to the computer industry. The theoretical approaches to metaphor in technical communication that may influence pedagogy can be further understood from the perspective of technical communication as a humanities discipline.

TECHNICAL COMMUNICATION THEORY

One of the stronger technical communication articles established metaphor as a topic of theoretical concern in 1984. In this article, Halloran and Bradford discuss what they refer to as the "anti-figurist tradition" as responsible for erecting barriers that inhibit the development of science. A little over twenty years ago, they too surveyed technical communication textbooks, where they found figures and tropes referred to as unnecessary displays of "erudition," a "literary trick," "mannerisms" or "tricks of meaning or word order" (1984, pp. 181-182). Halloran and Bradford assert that the plain style requires technical communicators to suppress their individuality since language is metaphoric (p. 182). They compare de Man's (1979) call for a language for philosophy that "comes to terms with the figurality of its language or . . . free(s) itself from figuration altogether" with Kinneavy's similar objections, noting that Kinneavy (1971) recognizes that "models and analogies are nonliteral terms necessary in science, so . . . the injunction cannot be absolute" (1984, p. 182).

When Halloran and Bradford (1984) explore the role that metaphor has played in the development of Watson and Crick's DNA theory (1953), they include the dead end reached when Crick, Griffith, and Orgel (1957) utilized the "comma-free code" metaphor, which theorized that the DNA contained no markers to separate distinct DNA units. Though he was incorrect, metaphor played a role in the falsification of the theory and contributed epistemologically.

Halloran and Bradford (1984) recommend a return to the teaching of figural language, by which they mean both figures and tropes, but they by no means recommend the return to the Renaissance educational experience. What they protest is the backlash that has relegated metaphor and analogy to the backwaters of technical communication where they still languish. They note that technical communication is "primarily visual" and discuss the role of lists in this respect. However, an argument can certainly be posed for technical communication as an aspect of visualization, as Wittgenstein (1958) has argued on language in general.

They recognize that some feel that the purpose of the technical writing course is to familiarize students with conventions, but that figures and tropes fall outside those conventions. It is also worth noting, though, that technical communication's rhetorical genres are, as models, analogies unto themselves. Halloran and Bradford add that teaching figures and tropes encourages students to question the field's conventions and to analyze audience to determine when those conventions might be effectively challenged and changed. Such an approach is certainly consistent with the intentions of this study.

More recently, J. Fahnestock (1999) has written of figures, which, though different from tropes, share the idea of a dependent structure with analogy. As a matter of fact, if one were to play word association with the phrase "figurative language," most people with an inkling of its meaning would think of terms such as "metaphor," "simile," and "personification." Though these are examples of what have been categorized variously as "figures of speech," they are tropes, which differ from other figures of speech in that a trope may depend upon the comparative aspect of a single word that in turn generates meaning, and in science, then shapes its conception. As examples, Fahnestock notes that words such as "current," "flow," "spark," and "discharge" evidence eighteenth-century indecision over whether electricity was "more like water or more like fire and firearms" (p. 5). Furthermore, figures of repetition such as ploche and polyptoton led to the development of the adjective "electrical," to the category noun "electrics," to the abstract noun "electricity," and finally to the verb "electrify." Fahnestock argues that what have traditionally been called the figures are also central to the development of scientific thought. In individual chapters, she examines the contribution of antithesis; incrementum and gradatio; antimetabole; and ploche and polyptoton.

After defining each figure or set of figures, Fahnestock (1999) traces the history of each, usually beginning with Aristotle's *Rhetoric* (1991) and discussing their theoretical development. Henry Peacham's (1954) sixteenth-century *The Garden of Eloquence* is a frequent touchstone, and so is Perelman and Olbrechts-Tyteca's (1969) *The New Rhetoric* for the twentieth century. Then Fahnestock discusses the use of the figure as an argumentative device. Finally, in the historical context, she illustrates how a variety of scientists have used the figure as an argumentative and epistemological tool.

Technical communication scholars can take a couple of points from Fahnestock (1999). Broadly, she calls for more historical research in this area; she has certainly not depleted the storehouse of figurative devices or their application in science. Specifically, she wonders if the figurative development of words to describe "electricity" is typical for scientific and technical terms.

In another early article that related metaphor to readability and pedagogy, S. M. Halloran and M. D. Whitburn (1982) view their discussion of metaphor through the lens of readability as a way of arriving at a plain style. They point to two problems inherent in readability in the early 1980s. First, educators such as K. W.

Hunt (1970) maintain that maturity in writing can best be measured by noting the length of the T-unit (an independent clause, plus any other embedded dependent clauses). However, readability formulas descending from the work of Flesch (1951) measure prose according to the length of sentences, so longer sentences cause a passage to receive a lower score. The current call for a plain style resonates with the same desire for a stylistic simplification apparent in seventeenth-century positivism that occurred partly as a rebellion against the classical education experience of the day. Halloran and Whitburn examine the Ciceronian approach that would synthesize a plain style (for instruction), a middle style (for pleasure), and a grand style (for persuasion) (1982, p. 61). However, even with the plain style, Cicero recommends that stylistic figures not call attention to themselves. Such a requirement does not suggest that they should be absent, though (p. 62).

Seventeenth-century science called for less human agency and greater reliance on instrumentation, which further removed the individual from scientific prose. Halloran and Whitburn quote the seventeenth-century scholar and popularizer of science, Bernard le Bovier de Fontanelle, who advocated geometry "as not so rigidly confined to geometry itself that it cannot be applied to other branches of knowledge as well" (1982, p. 66). Halloran and Whitburn (1982) compare such an approach to those who would quantify prose with readability formulas. Furthermore, such excesses place into the quantifying camp those who would actively crusade against figures and metaphor in technical communication. The result is that students who are not taught to use them are left to their own devices if they are to learn them at all.

Textbooks in general are short on explanations of metaphors and long on examples. As a result, students learn from imitation and induction, not through a theoretical approach. Halloran and Whitburn (1982) further depict such learning as a fallacy by contrasting it with how scientists learn the craft of science, not by studying only its theory but by practicing it in the laboratory, under supervision. Furthermore, scientific practice tends to divide and classify, with more emphasis on the part than the whole. Metaphor in scientific thinking provides a unity to thought and expression.

Another good example of an early study of metaphor in technical communication is Whitburn, Davis, Higgins, Oates, and Spurgeon's (1999) article "The Plain Style in Technical Writing," which first appeared in a 1978 issue of the *Journal of Technical Writing and Communication*. Editors T. C. Kynell and M. G. Moran picked the article as what they refer to as a "Landmark Essay" for the their book *Three Keys to the Past: The History of Technical Communication*. In this essay, Whitburn et al. conclude, "the past . . . holds stylistic riches for the modern practice of scientific and technical writing" (p. 130). Such a conclusion is supported by examining the abrupt point of change in scientific writing that occurred with the seventeenth-century scientific revolution. Prior to this paradigm shift, scientists revealed knowledge through scholarly work derived from reading, writing, and reasoning. The scientific revolution changed how scientists acquired knowledge through observation, so "words, associated with the rejected science of the past, became suspect" (p. 125).

Whitburn et al. (1999) note Thomas Sprat's attitude toward scholarly writing. Sprat wrote that "rhetorical ornamentation" stood "in open defiance against Reason: professing not to hold much correspondence with that; but with its Slaves, the Passions . . ." (p. 125). The problem, according to Whitburn et al., is that revolutions typically occur in response to overindulgence, and the revolution itself does not so much serve as a corrective as it does an impetus to swing wildly in the opposite direction. Whitburn et al. argue that the plain style advocated today in science is another extreme. As a result,

> Scientific or technical students rarely confront a writing task armed with sufficient stylistic tools to shape their compositions aggressively. They tend to be more concerned with what *not* to do than what *to* do in their writing [Whitburn et al.'s italics]. Scientific and technical writing textbooks tend not to drill students in even such basic stylistic techniques as antithesis, climax, parenthesis, and apposition. On the contrary, students are expected to muddle their way through by stylistic instinct. Such an approach promotes the myth of the born writer (1999, p. 126).

In this sense, the traditional pre-scientific revolution approach to writing represents a rhetorical strategy vanquished, and such loss is a twentieth-century phenomenon since nineteenth-century scientists were still being drilled in traditional approaches to tropes and figures. Whitburn et al. note that prior to the seventeenth century, scientists were educated in the classical tradition "to vary a theme hundreds of different ways . . . through comparison, example, description, repetition, periphrasis, and digression" (1999, p. 126). During the Renaissance, students might learn over three hundred different figures of speech and be expected to know how to use them. Indeed, scientists through the nineteenth century were educated classically, more often than not to their chagrin. Whitburn et al. do not suggest a return to such an approach, but they do suggest that such extreme swings in attitudes toward writing may represent the unnecessary exclusion of what may otherwise serve as a supplement to a scientist's rhetorical repertoire.

Whitburn et al. (1999) also consider figures as a shortcut. They cite Erasmus, who posed that conciseness can best be achieved through familiarity with a number of words and figures of speech. However, they note that scientists and engineers lack the training that would allow them to use figures effectively, even going so far as to suggest "irony, hyperbole . . . litotes . . . again one searches in vain for information about such devices in technical writing textbooks," a fault they attribute to adherence to the plain style (p. 127).

Just as today the value of wind energy, a technology with roots in ancient Persia, is beginning to be realized again, so too should we consider the value of metaphor in scientific and technical communication. Whitburn et al. (1999) point to teaching metaphor in the technical communication classroom, and their connection with theory is stronger, in a sense, but what theory we are being drawn to is important to consider. What is perhaps strongest in Whitburn et al. is a sense of composition theory, that of nullifying the idea of the immortal born-writer that creates so much anxiety for the mortal ones. While Whitburn et al. lend a sense of the problem's history, the ties to theory are weak. For example, while it is interesting to know that seventeenth-century scientists were required to "to vary a theme hundreds of different ways" (p. 126), why, then, has metaphor been left out of the education of not just scientists, but really all students, other than pointing out tropes in the literature classroom? Education had not greatly changed 300 years later in the nineteenth century, so the omission of metaphor from the scientist's training is more of a twentieth-century phenomenon. Charles Darwin's education, for example, was quite the classical one. Indeed, the relegation of metaphor to the literature class is similar to rhetoric being relegated to the public speaking course. The problem can be read as twofold: first, science was not the focus of an education and in many cases was not taught at all through the nineteenth century; second, the traditional interpretation of Aristotle's *Rhetoric* (1991) encouraged the teaching of metaphor as a noun, or an object, rather than as a verb capable of transforming science, as Ricoeur (1975) has noted and as I shall discuss in further detail in the next chapter.

Not all of the early essays on metaphor contributed significantly to the development of a theory of metaphor in technical communication, and some perhaps discredited metaphor to a degree. Part of the problem stems from training in literature studies. Though I would be the last to decry the presence of the humanities in technical communication, when a scholar with a heavier investment in the traditional humanities approaches technical communication, the result is often skewed in the direction of the scholar's academic background. Since most technical communication courses were initially taught in departments of English (and those that were not were more likely to be taught by someone with a scholarly background in English), theory is often derived to some extent from literature studies. Though such comment enriches technical communication and provides a humanistic perspective that the field will always need, literature studies does not provide as strong a background as rhetorical theory. Good examples are J. S. Harris' (1975, 1986, 1993) articles. First, Harris should be commended for not only being the first to pose the subject of metaphor for discussion in technical communication, but for persisting in its discussion for almost 20 years. The problem is that all of the articles are largely inductive, which points to how metaphor has been discussed in scientific and technical communication and how it has been presented in technical communication textbooks. In the 1975 article, Harris cites numerous examples of metaphors from life sciences, physical science, engineering, and technology. For technical metaphors, for example, he lists 17 of the 26 letters of the English alphabet, from A-frame to Z-section.

Harris' (1975) only approach to rhetorical theory in the initial article is with the issue of dead metaphor. He points out that "manufacturing" originally meant to create an object manually. Eventually, the term more generically meant simply "to make," but today, "non-manufactured" goods have been created manually. As a result, according to Harris, the term "non-manufactured" has been "re-metaphorized." The point, for Harris, is that metaphor in technical communication is not endowed with the "transcendency . . . [of] poetry" (p. 11). Though the observation of the "re-metaphorizing" of "non-manufactured" is interesting, some knowledge of theory would allow more insight into the process Harris has observed. Harris is not clear in terms of what he means by the "transcendency" of poetry, but it is fair to question the extent that he considers the epistemological significance of metaphor. For example, Perelman and Olbrechts-Tyteca (1969) would comment that this shift in meaning is part of the process that a metaphor passes through as it exhibits what McMullan (1976) has named fertility. While Harris does not condemn metaphor on the basis of the shift of meaning, neither does he praise it.

What would technical communication scholars who are unconvinced of the value of metaphor to technical communication pedagogy make of the slippery nature of "manufacturing," especially if they are skeptical of the value of teaching metaphor as a part of a technical communication course? It would not be unreasonable to expect them to ignore such a discussion. That metaphor in technical communication lacks "transcendency" depends upon the type of work the student might eventually engage in. Whether a student becomes a scientist, an engineer, or a technologist, metaphor can play a generative role in epistemo-logical expansion and communication. Though Harris does well to point technical communication in the direction of metaphor, he does not relate it enough to theory to lend it credibility.

Harris' other two articles are even more inductive (1986, 1993). One article (1986) catalogs different types of shape metaphors in technical communication. He describes how shape imagery is drawn from geometry and nature, which ranges from aspects of the human form to other animals and to plants and other figures drawn from this wellspring of influence.

The third Harris article (1993) is also inductive in the sense that it consists largely of his poetry. Harris' most significant allusion to theory in this essay concerns paralleling mathematical equations with metaphor. Unfortunately, this discussion exists in a theoretical vacuum. Harris is evidently unaware of Hesse's (1970) dialogue between the Duhemist and the Campbellian. The Duhemist's reliance on the mathematical model becomes a flaw in the argument against metaphor. Loewenberg (1973), too, has argued that the mathematical renders a model that may be too closed to interpretation. McMullan (1968) would point out that the model leads an existence unto itself, separate from the mathematical model that is still only a model. And of course, Lakoff and Nunez's (2000) work on mathematics as metaphor that is explored in chapter four would also well inform this discussion.

Harris (1993) also cites a "recent article" in *College English* whose author "claimed to have searched the literature and found no mention of metaphor in technical writing texts" (p. 313). Harris' essay was originally presented as a 1990 Barker Lecture at Brigham Young University. He may be referring to a Jerome Bump (1985) article since it is the first article on the topic of metaphor in technical communication to be rendered from an examination of the journal prior to 1993. Bump's essay is discussed later in this chapter in greater detail as it relates to the technical communication classroom.

Though Harris does well in these three articles to point technical communication in the direction of metaphor, he does so with little or no theoretical direction. Reference to theory seems derived largely from literature studies. Including the theoretical background from this study, on the other hand, would enrich Harris' perspective and provide greater impetus for inquiry and teaching in this area. Without such focus, Harris' recommendations seem little more than eccentricity, rather than what would warrant a pedagogical response grounded in theory.

Another problematic seminal article is J. R. Gould's 1979 essay that argues that literature studies in the traditional English department maintain that technical communication as an academic field cannot develop meaningful theory. If technical communication cannot travel beyond practical application, then Gould concurs. In other words, Gould was content to meet the English departments on their own terms. The development of computers in general and usability studies in particular have shown that Gould did not anticipate the future path of technical communication. Still, a way to meet the traditional English department on its own terms and reach beyond the practical is to apply rhetorical theory to the history of technical communication as a way of arriving at theory. A number of technical communication scholars have foraged in this field, with a particular focus on metaphor. One problem is with those who stopped with Francis Bacon and accepted his writings as a mandate for a plain style.

Many scholars have cited Francis Bacon as the source of what is thought of as the plain style in scientific and technical communication (Moran, 1985, pp. 28-30). Christopher Baker (1983) illustrates why Francis Bacon, whose influence is felt today in technical communication, is generally regarded as representing a turning point in scientific writing. Baker points out that Bacon can be credited with reforming scientific writing into "a technology of style, a theory of communication designed specifically for the transmission of scientific fact" (p. 118). One of Baker's intents with this essay seems to be to reconcile metaphor with scientific and technical writing. For example, Baker quotes Bacon, who wrote that knowledge "is delivered as a thread to be spun on [and] ought to be delivered and intimated, if it were possible, in the same method wherein it was invented" (as cited in Baker, 1983, pp. 119-120). Despite the simile, and Bacon's writing is rife with them, Bacon felt there should be a one-to-one correspondence between the word and the object or idea, which suggests what is now thought of as the "plain style in scientific and technical communication." Conversely, Bacon advised science writers who would communicate with a general, educated audience to use metaphor. Though contemporary technical writers strive, as Bacon did, for conciseness, objectivity, and topical organization, they should remember that Bacon did not eliminate metaphor as a tool.

Carol Lipson (1985), too, would point to a reevaluation of Bacon. Bacon's advice on scientific writing is so riddled with contradictions that Lipson delineates them as a way to deconstruct him. She would rather think of him as the first deconstructionist than as the father of modern scientific writing, as J. P. Zappen (1999, 1975) and others have hailed him. However, Lipson bases her deconstruction on the way Bacon advises science writers to approach the scientific community and the more general audience that Bacon characterizes as those of "vulgar opinions" (p. 150). Lipson points out that for this reason, the Royal Society devalued metaphor. However, she agrees with Baker (1983) that Bacon valued metaphor for communicating with a general audience.

It is worth noting Bacon's general disdain for Aristotle expressed in *The Advancement of Learning* where Bacon (1952) discredits him as the source in a "degenerate" classical education that

> did chiefly reign amongst the Schoolmen: who having sharp and strong wits, and abundance of leisure, and small variety of reading, but their wits being shut up in the cells of a few authors (chiefly Aristotle their dictator) as their persons were shut up in the cells of monasteries and colleges, and knowing little history, either of nature or time, did out of no great quantity of matter and infinite agitation of wit spin out unto those laborious webs of learning which are extant in their books (p. 12).

The treating of metaphor as an object (or noun) is endemic to classical rhetoric, stretching back to Cicero and the author of the *Rhetorica ad Herrenium*. Bacon is more likely to read Aristotle in this vein as well, so his admission of "similitudes" as valuable to scientific communication is probably one he did not think of as Aristotelian.

Since it is often historical, rhetoric of science is somewhat related to these types of concerns, and again, some technical communication scholars have focused on metaphor. R. Johnson-Sheehan has written more specifically of metaphor in physics (1995, 1997). In his essay on the *Special Relativity Theory* (1995), he identifies a number of metaphors. Among them, the theory of relativity becomes the primary metaphor that altered physics and geometry as other scientists adopted the metaphor. It also transformed what were considered to be dead metaphors such as "space" and "time." As a result, some dead metaphors such as that of the luminiferous aether were dispensed with as they became obsolete. The operative metaphor of relativity became a defining moment for a paradigm shift.

Johnson-Sheehan's (1995, 1997, 1998) publications in this area thus far have focused more on the science than on the metaphor. He mentions theory in passing, but his work is largely inductive and would benefit from greater attention to theory. For example, though he relates that the aether was discarded as a result of shifting metaphors, he does not connect the paradigm shift with theory. To what extent does falsification interplay with metaphor? Johnson-Sheehan demonstrates rather than analyzes.

He has also used his method of "metaphorical analysis to determine whether or not Max Planck invented the quantum postulate" (1997, p. 177). Johnson-Sheehan rightly points out that what is perhaps most rhetorically significant is the way that Planck layered the metaphors of "energy spectrum" and "entropy" to create the metaphor of "black body radiation" (1997, pp. 183-184). However, when he discusses theory, Johnson-Sheehan introduces Kenneth Burke as "probably the first to recognize the relationship between metaphor and rhetorical invention in science," which ignores the contributions of Nietzsche, Richards, and, as I shall discuss in the next chapter, many others. Johnson-Sheehan mentions briefly the work of Black, Hesse, Kuhn, and others, but he privileges Burke's work, which is much less focused and less detailed than Hesse or Black's. Any of these authors has much more to say than Burke on the subject. Johnson-Sheehan then poses, "So, instead of offering yet another oration on the importance of metaphor to science, let me move ahead by discussing what metaphors *do* [Johnson-Sheehan's emphasis] in scientific discourse," (1997, pp. 178-179), which amounts to chronologically reversing the development of the theory of light, beginning with Young and touching upon Newton and Descartes. However, given the contribution of Johnson-Sheehan's essay, that metaphors were layered to create metaphor, his explication would have benefited from including some discussion of the interactive quality of metaphor. Where does his interpretation fit in the theoretical dialogue? Indeed, considering the interactive quality of metaphor would, at the least, augment rhetorical theory of metaphor. It would certainly not be merely "yet another oration on the importance of metaphor to science," and it could yield a new light on Johnson-Sheehan's interpretation of Planck's theory. The result would strengthen theoretical weakness in his essay.

More recently, Johnson-Sheehan's work (1999) has taken a more interesting theoretical turn. He has advocated metaphor as a hermeneutic device manifested in narrative created by the context the metaphor creates, and he alludes to science as narrative.

Another author representing the rhetoric of science is J. E. Harmon (1986, 1994), who notes that though scientific and technical writing is supposed to be plain, the scientist often makes use of a variety of figures and tropes. An extreme example includes Galileo's use of an anagram to establish his discovery of Saturn's rings as a way of validating when he discovered them, much as a scientist today would publish a note in a journal. For a more recent example, Harmon (1994) examines the role of analogy in a 1923 thermodynamics textbook that compares the structure of knowledge in thermodynamics to a cathedral. Harmon notes that in a work on thermodynamics, J. D. van der Waals stated that, "the motion of the planets and the music of the spheres will be forgotten for awhile in admiration of the delicate and artful web formed by the orbits of those invisible atoms" (as cited in Harmon 1994, p. 314). He also discusses the metaphorical role of neologisms as they relate to the development of science.

Harmon has proposed that metaphor can "convey information inexpressible, or at least not easily communicable, by ordinary language." In particular, he is impressed with metaphor's conciseness and precision (1986, p. 180). Harmon notes that metaphor can be a conduit for colorful imagery in work that would otherwise lack it. Tracing the path a metaphor might follow, Harmon has observed how metaphoric terms spread from the scientific to a lay audience (1986). However, he (1994) notes that in his study of the 89 scientific documents culled "from Eugene Garfield's top 400 most-cited documents in the *Science Citation Index*" (1945-1988), which he examined for their use of metaphor, the writers infrequently used metaphors. Though metaphors can claim historical significance, their appearance is quite rare, according to Harmon. The reason may be that scientific writing aims to report, so the scientist focuses on accuracy rather than fanfare. Harmon recognizes that metaphor is regarded by many in the scientific community as verbal sleight of hand. Therefore, even if a scientist were to invoke a metaphor, an editor may wean it out.

As with Johnson-Sheehan, Harmon's approach (1986, 1994) is largely inductive. Though his study contributes to the tracing of a metaphor from the scientist to the lay audience, his work would benefit from including rhetorical theory. For example, when he writes of identifying what he refers to as "poetical metaphorical language," he observes that in 89 papers from the *Science Citation Index* (1945-1988), he identified 42 "poetic metaphors" in 21 papers (1994, p. 189), which flies in the face of rhetorical theorists from Nietzsche to the present, who maintain that all language is metaphoric. Harmon admits that determining what is and is not metaphoric is somewhat subjective, and he claims to determine metaphoric use through assigning terms to categories of standard usage versus usage that is not standard, or metaphoric, but he does not reveal how he decides which terms belong in which category. Why is "messenger RNA" (1994, p. 187) an original metaphor when "regulator and operator" is not (1994, p. 190)? Further exploration could explain more about how and why scientists use metaphors as well as how they could be used more effectively. Of course, how and why scientists use metaphors can be influenced by how they are taught to use them, so it is next useful to consider how metaphor is being taught in technical communication.

TECHNICAL COMMUNICATION PEDAGOGY

Though pedagogical concerns permeate the essays I have considered so far, others specifically address this concern. J. Bump (1985) surveyed the technical communication terrain and found it lacking in the way that it addressed metaphor and analogy in textbooks. Bump excoriates technical communication textbook authors for the paucity of information on metaphor because there, it was only "conspicuous by its absence" (1985, p. 444). Like Whitburn et al. (1999), Bump blames the problem on the concept of objectivity in the scientific discourse community. He introduces this idea by discussing the negative connotation of personification, which he sees as a problem for proponents of metaphor in scientific and technical writing because "our tendency to personify . . . suggests the introduction of personal bias and emotion into science" (1985, p. 446). Bump's assertion is an apt one since scientists would consider objectivity to be an aspect of basic scientific competence and ethics, but it flies in the face of the extensive personification used to describe scientific processes such as cloning.

V. M. Winkler (1983) ties composition theory on process approach with technical communication and metaphor. According to Winkler, technical communication has long been aware of audience analysis, but without fully exploiting invention, as composition does. Instead, models are used extensively in science for invention. Typically, science's models are organizational, and analogy can certainly be thought of as a figure or a scheme more readily than a trope or metaphor. Regardless, like a metaphor, an analogy transmogrifies the abstraction to the concrete. She draws a parallel between Leatherdale (1974), and Young, Becker, and Pike's (1970) four stages that the writer engages to solve rhetorical problems. Winkler notes that Young, Becker, and Pike's process of preparation, incubation, illumination, and verification can be equated with Leatherdale's extended definition of the interaction of analogy. She then ties these ideas in with Burke's pentad (1969) of act, agent, agency, scene, and purpose, which can be thought of as "imported analogies" (1983 p. 119).

The suitability of the technical communication scholar to teach metaphor has been supported as well. On the issue of competence, K. N. Hull (1980) has noted that an instructor with a background in literature is more likely to recognize metaphor when a student uses one, and the instructor is qualified to nurture its use. Hull recalls spotting metaphor in student work, but he cannot help but "doubt if the writer was aware of the fact since so few students really are taught the devices of classical rhetoric" (p. 881). The technical communication classroom certainly affords the opportunity for teaching such an approach to scientific and technical thought, one that is supported by the scholarly literature.

Just as with the other articles, there are some problems with the pedagogy essays. Certainly some of Russell Rutter's (1985) thoughts on metaphor in technical communication are valuable. For example, on the basis of the intellectual involvement with creative prose that nonetheless models good technical communication practices, he advocates the traditional literature scholar for teaching technical communication courses. He mentions M. Halloran's article (1997) that explicates Watson and Crick's metaphorical presentation of DNA, but Rutter himself does not utter the name of metaphor. This section of the article concludes with a paragraph wherein he mentions Smeaton's description of the processes he underwent to conceive the structure of the Eddystone Lighthouse as "imaginative analogy" before moving to another example to conclude the section. In his conclusion, he reminds instructors with humanities backgrounds that "they possess a vital insight, that all communication is an imaginative projection of concepts onto otherwise meaningless data to produce orderly, informative statements that satisfy particular needs" (1985, p. 709), again without mentioning the dread name of metaphor.

Patrick Moore's (1996) call for technical communication scholars to differentiate between what he refers to as an instrumental versus the traditional rhetorical approach, which could be read as a refutation of metaphor in technical communication, is interesting to consider. Moore argues that technical communication scholars such as Dobrin (1983) and Miller (1979) have advocated a more humanistic technical communication that includes literary theory as an approach. Countering with the concept of what he calls "instrumental discourse," Moore argues that, positivism aside, technical communication is a specific type of communication for a specific purpose. Clear communication that saves or improves lives should not be strained through a theoretical filter that would put graduates of technical communication programs at odds with communicating clearly. It is more ethical, according to Moore, to agree that technical communication is instrumental in the sense of words having a particular meaning than to wonder to what extent language is being flattened by striving for an unobtainable objectivity. Technical communication, Moore asserts, is a social construction.

To start, the idea of creating an instrumental language is not new; I. A. Richards (1943) professed the same goal and met with failure. Language, the human experience, and what Perelman and Olbrechts-Tyteca (1969) would call the universal audience demand too much from the technical communicator. Examining the role of metaphor in technical communication should not be read as applying the tools of literature to humanize technical communication. Rather, this book examines metaphor as a discredited tool that some scientists have laid by the wayside, mistaking it for an anachronism. Furthermore, though all language as metaphoric is a cornerstone of this book's argument, the use of metaphor and analogy is a particular application of language that is somewhat instrumental without falling into the trap of considering metaphor as an object. In addition, this book recommends theory drawn from rhetoric and philosophy, not literature studies. That the nature of metaphor has been explored theoretically from these perspectives encourages an approach that is not at all bound by literary theory. Rather, aligning rhetorical theory with the philosophy of science aligns with the aims and goals of the technical communicator rather than the literary critic.

CONCLUSION

Metaphor emerges as a rhetorical tool that continues to interest technical communication scholars. From the beginning, scholarly approaches to metaphor in technical communication have been largely inductive, which points more to the background for metaphor studies experienced by many technical communication scholars. Most people identify metaphor with the literature classroom where tropes and figures are experienced and explained as they serve as a motif to support a theme, at best, or as mere decoration, at worst. Today the development of artificial intelligence has encouraged the development of metaphor studies, a cross disciplinary field drawn from rhetoric, philosophy, linguistics, psychology, and education. A literature has developed that would well inform the scholarly technical communication community and further justify the inclusion of metaphor in technical communication pedagogy. Metaphor, as the examination of the textbooks in the first chapter indicates, is encouraged, but not wholeheartedly. Technical communication is still haunted by the idea of a plain style as the preferred, as Bacon casts his long shadow over the field. Technical communication approaches to developing metaphor theory are rife with problems and lean toward oversimplification. The result is that the interactive aspect of metaphor and its epistemological consequences are ignored. Few questions are asked in technical communication of how metaphor may be used to communicate, much less how to create new knowledge. The next chapter surveys metaphor theory from rhetoric, philosophy, and linguistics to create justification for applying metaphor theory to technical communication pedagogy; the chapters following it demonstrate how metaphor theory could direct the use of metaphor in technical communication.

CHAPTER 3

A Review of the Theories
of Metaphor

A variety of perspectives could allow us to discuss metaphor in technical communication. Many studies of the power and uses of metaphor have been carried out in psychology and in education, especially those that explore how metaphor aids, or interferes with, reading and other cognitive processes. The rhetorical perspective is an appropriate focus for metaphor because of the role that rhetoric has played in the shaping of academic technical communication programs. Indeed, rhetoric has been instrumental in shaping technical communication programs, which have adopted the rhetorical canons, especially invention, arrangement, style, and delivery. Metaphor is a stylistic element, so why is technical communication's perspective on it somewhat uneven? It could be argued that science has embraced metaphor, but this book will offer evidence that the use of metaphor in science is largely unconscious, rather than self-conscious, as it should be if it were used more effectively.

Metaphor can be discussed in other contexts: literary, rhetorical, and philosophical. To achieve coherence for technical communication, this chapter limits itself to those works that provide a unifying thread among those disciplines. Rhetoric and philosophy are discussed concurrently; that dialogue then provides a background for a later discussion of metaphor in technical communication. The discussion centers on how theory has shaped the conception of metaphor. For that reason, I approach it by considering first the substitution view, which began with a misinterpretation of Aristotle, who cast metaphor as worthy of philosophy. However, I consider Aristotle's influence on twentieth-century rhetoricians as well. Next, I proceed to the interaction theory and then to metaphor as epistemology.

First, the matter of considering philosophy and rhetoric concurrently should be justified. Such justification can be traced to the changes Newton wrought upon science when he pushed probability to the forefront of scientific thought. When Newton empowered probability, he devalued the syllogism and increased the power of the enthymeme. For science, Aristotle (1952a) privileged the syllogism, relegating the enthymeme to rhetoric. However, when Newton sanctioned probability, he elevated rhetoric by espousing the hypothetico-deductive as science's methodology, which moved science from the universalistic tradition of Aristotle that dealt with certainties (hence syllogistic reasoning) to something akin to the current scientific method allowing probabilities. As a result, rhetoric became more meaningful to scientists, and metaphor traveled as part of the rhetorical baggage, flourishing in scientific style.

Though Newton empowered the rhetorical, those who would devalue it found a champion in Francis Bacon, who called for a use of language in science writing that has since then been thought of as the "Plain Style," and that has been interpreted as calling for an end to the use of metaphor in science writing. Actually, Bacon's discomfort with florid style can just as easily be traced to syntax as to metaphor. Specifically on metaphor, Bacon's attitude is a true dichotomy. In Book I of "The Advancement of Learning," he warns against a time when

> men began to hunt more after words than matter; more after the choiceness of the phrase, and the round and clean composition of the sentence, and the sweet falling of the clauses, and the varying and illustration of their works with tropes and figures, than after the weight of matter, worth of subject, soundness of argument, life of invention or depth of judgement (1952, pp. 11-12).

This passage is often cited as evidence of Bacon championing a plain style in scientific and technical communication. It is not commonly recognized in technical communication scholarship that Bacon also valued metaphor, for he also notes in Book II of "The Advancement of Learning," that those who write on scientific subjects, especially for the general public

> have a double labour; the one to make themselves conceived, and the other to prove and demonstrate. So that it is of necessity with them to have recourse to similitude and translations to express themselves . . . for it is a rule, that whatsoever science is not consonant to presuppositions must pray in aid of similitudes" (as cited in Baker, 1983, p. 122).

Metaphor did not disappear with Bacon, not even after Thomas Sprat, who in his *History of the Royal Society*, encouraged scientists

> to reject all the amplifications, digressions, and swellings of style: to return back to the primitive purity, and shortness, when men deliver'd so many *things*, almost in an equal number of *words* [Sprat's italics]. They have

exacted from all their members, a close, naked, natural way of speaking; positive expressions; clear senses; a native easiness: bringing all things as near the Mathematical plainness, as they can: and preferring the language of Artizans, Countrymen, and Merchants, before that, of Wits, or Scholars (1958, p. 113).

It is worth noting in this passage the irony of Sprat's simile as he calls for "all things as near the Mathematical plainness" in scientific prose.

Metaphor is a necessary component of human thought. As I shall show, many rhetoricians and philosophers have proclaimed all language as metaphoric, and it is especially important for scientific thinking, where it serves a generative purpose, as the case studies demonstrate. First, I examine relevant metaphor theory.

Metaphor did not disappear from scientific and technical communication after Bacon and in fact persists to this day. Arguments against metaphor in scientific and technical communication are largely a twentieth-century phenomenon, and the construction of atomic theory is an important landmark in metaphor's decline in scientific writing, one, too, that persists to this day and that is well exemplified in the controversy surrounding cloning, where no metaphor has surfaced to communicate scientists' aims to the public. With these thoughts in mind as a grounding, the theoretical literature should now be examined.

SUBSTITUTION THEORY OF METAPHOR

The substitution theory may be defined as when one word is thought to be substituted for another to create a comparison that is understood metaphorically, or figuratively, rather than literally. With a metaphor such as, "An atom is a miniature solar system," the "atom" and the "solar system," as nouns, are treated as objects that one may substitute for the other. In a sense, there is a tendency to treat the metaphor's elements as objects. Such a comparison does not suggest how or why the metaphor works. Instead, it tends to cause any conscious usage to focus on the different types of tropes rather than on how they work and how they might be useful. As this chapter demonstrates, Aristotle was the first to consider metaphor as a topic worthy of discussion. However, classical scholars and others misinterpreted his work as what became known as the substitutionist approach, a perspective that considers metaphor as ornamental. Such was not Aristotle's intent.

ARISTOTLE ON METAPHOR

Although many philosophers and rhetoricians have written of metaphor, Aristotle first recognized its qualities as worthy of philosophy. That the nineteenth-century scientists discussed in Chapter Four were classically educated and would have been familiar with his rhetoric is another reason to begin discussion of metaphor and analogy with him.

Aristotle's *Rhetoric* (1991) is the basis for thinking about metaphor from rhetorical and philosophical perspectives, and it takes the shape that all grade school students are familiar with from the analogies of standardized testing, of A is to B as C is to D, which Aristotle himself models as A:B::C:D. He also describes metaphor as A:B. In this way, Aristotle links metaphor with analogy, just as they are linked in this book. Furthermore, of metaphor, he posits that it "most brings about learning," and that "the greatest thing is to be a master of metaphor. It is the one thing that cannot be learnt from others; and it is also a sign of genius, since a good metaphor implies an intuitive perception of the similarity in dissimilars" (1952b, p. 694). Since metaphor "most brings about learning," then it can said to be epistemological, and it has the potential to be generative.

Aristotle defines metaphor by specifying how it generates "transference . . . either from genus to species or from species to genus or from species to species or on the grounds of analogy" (1952b, p. 693). Though this definition is garnered from *On Poetics*, Aristotle refers readers to it often in *On Rhetoric,* and once specifically in his discussion of metaphor (1991, p. 223). In his definition of metaphor in *On Rhetoric*, Aristotle mentions analogy, but he continues his discussion of analogy later in Chapter Ten.

Aristotle's explication of analogy is one of the most important ways that he extends the complexity of his discussion of figurative language. Metaphor is not simply a matter of comparison, of explaining the known in terms of the unknown, which Black (1962) also called the comparison theory of metaphor. Similes differ from metaphors because "like" or "as" explicitly expresses the comparison, and Aristotle likens them to "metaphors needing an explanatory word" (1991, p. 230). He writes, "of the four kinds of metaphor [the fourth is epithet], those by analogy are most well liked" (1991, p. 246). In *On Poetics,* he specifies, "analogy is possible whenever there are four terms so related that the second (B) is to the first (A), as the fourth (D) to the third (C)" (1952b, p. 693). This expression of "shaped language" (1991, p. 245), as he calls it, is related to his concept of the syllogism since it depends on premises and conclusions.

Into his discussion of metaphor, Aristotle brings his idea of enthymeme, whose point, he says, is to "create quick learning in our minds" (1991, p. 245). Roughly classifying ineffective and effective enthymemes, he categorizes the first type of ineffective enthymeme as "superficial," which he defines as "those that are altogether clear and of which there is no need to ponder"; the second type of ineffective enthymeme as that which is "unintelligible"; and the effective enthymeme as "those . . . of which there is either immediate understanding when they are spoken, even if that was not previously existing, or the thought follows soon after; for [then] some kind of learning takes place" (1991, p. 245). Aristotle more specifically links enthymeme with shaped language, or metaphor, by noting that "such kinds of enthymemes are well liked; in terms of the *lexis,* [an expression is urbane] . . . because of shaped language" (1991, p. 245). Enthymeme, then, can be tied to metaphor, which incorporates it into Aristotle's rhetoric.

Of these three categories, e-mail would fit as a metaphor with superficial clarity; it does not provide any epistemological insight, other than as an operative metaphor for its user. The Solar System Analogy, however, allowed its users an immediate understanding that then left them to further speculate on the structure of the atom.

A metaphor is similar to an enthymeme in that an enthymeme is a truncated syllogism, and a metaphor is a condensed analogy. For example, the SSA can be presented in a number of different forms: as a metaphor: "An atom is a miniature solar system; as a simile: An atom is like a solar system; or as an analogy: "As the planets orbit the sun, subatomic particles orbit the atom's nucleus." The analogy can be further extended: "When two metals are fused, it is similar at the atomic level to when a comet joins the solar system." The metaphor's first part, A:B, is assumed to be true because it is within the audience's realm of experience. With that assumption, it can be concluded that the same relationship exists in C:D. Or as it might be expressed analogously, an enthymeme is to a metaphor as a syllogism is to an analogy. In either case, metaphor and enthymeme ask the audience to provide a context, what Perelman and Olbrechts-Tyteca (1969) call the "theme" (p. 369), or what Johnson-Sheehan (1999) refers to as the audience as the "inventor of meaning for a metaphor" (p. 60). As an example, Aristotle quotes Pericles, who said that "the young manhood killed in the war vanished from the city as though someone took the spring from the year" (1991, p. 246). As in other cases of Aristotle's examples of analogy, the A:B relationship is not as clear as it would be on a standardized test. However, Aristotle presents another example: "Thus a cup (B) is in relation to Dionysus (A) what a shield (D) is to Ares (C)" (1952b, p. 693). Although this example lacks the metaphoric element of Pericles' spring being taken from the year, it highlights the comparative aspect of analogy. In Pericles' analogy, another element is present: that "strangeness" Aristotle alludes to that is expressed by metaphor. The lack of a clear A:B::C:D relationship is similar to the assumed premises of an enthymeme. Pericles' metaphoric element provides movement from the literal deaths of young men into a completely imaginary realm without a spring that is nonetheless meaningful. The analogy drawn between Dionysus' cup and Ares' shield differs in that as a model, it is too flat to convey the metaphoric element of epiphany that can be equated with knowledge.

Engineering students are generally familiar with such metaphors. On occasion, professors in their discipline may use similes such as, "Current in a wire is like water in a hose." In this case, the A:B::C:D relationship is clear, and it provides a model of thinking for students. How often, however, do students ask the professor about the nuances of the simile? How often does the professor explain those nuances? The technical communication classroom is the place for that dialectic with a professor who has the background knowledge to discuss and explain it.

On the difference between metaphor and simple comparison, a number of more contemporary rhetoricians and philosophers have weighed in. Consider this definition of the Doppler Effect (1998): "A change in the frequency with which waves (sound, light, radio) from a given source reach an observer when the source and the observer are in rapid motion with respect to each other so that the frequency increases or decreases according to the speed at which the distance is decreasing or increasing." Or is it clearer to illustrate it with the way that the sound of a siren seems to change pitch as it approaches a stationary observer? In this case, the explanation with the changing pitch of the siren is a comparison that lacks the metaphoric element. It can create knowledge for the audience, and that knowledge can be directly applied to understanding a physical phenomenon, but it is clearer as a model than as a metaphor. It is a model in that it represents a phenomenon, but it is not useful as a metaphor because its epistemology is limited to explaining a particular case rather than generating theory.

The Doppler Effect model leads to another dilemma: this book argues that all language is metaphoric. If so, then how can there be a distinction between model and metaphor, or between those concepts and language in general? The answer lies in the idea of self-conscious usage that is aimed toward generating knowledge. When a scientist poses "An atom is a miniature solar system," and then proceeds to use the metaphor to generate theory, the self-conscious usage is epistemological. When the professor poses for a class that "Current in a wire is like water in a hose," the simile is epistemological for the student but not for the professor. All language is metaphorical, but this book is chiefly concerned with the role of the epistemological metaphor.

Perelman and Olbrechts-Tyteca (1969) have also argued that if comparisons "belong to the same sphere," then no metaphor or analogy has taken place. It is in that case a comparison (p. 373). With the Doppler Effect, while the comparison is effective, it is not metaphoric since the general idea of sound waves is compared to a specific instance of a type of sound wave—the siren. To what extent can knowledge be created when a more fertile metaphor is applied?

The link between metaphor and analogy with both enthymeme and syllogism is important because metaphor and analogy are linked to persuasion and to the epistemological. With the link to the syllogism, metaphor is linked to the logical underpinnings, especially in Aristotelian thought, of scientific thought and expression. With the syllogism, the premises must be held to be true for the syllogism to function mechanically, and the enthymeme is but an abbreviated syllogism.

Though he did not think it could be taught, Aristotle highly valued metaphor, believing it to be a mark of genius. The point at which he philosophizes it is when he discusses the idea of motion associated with metaphor, the concept of *energia*. Such movement provides the occasion for an epiphany through the identification of the similarities that allow arrival at a new state of knowledge we think of as a Kuhnian paradigm shift. This awareness is at the heart of *energia*, which is related to his concept of urbanity, and it is with urbanity that he arrives at a philosophical touchstone for "shaped language," or *schemata,* as he concludes with "thus . . . urbanities come from metaphor by analogy and by bringing-before-the-eyes" (1991, p. 248). "Bringing-before-the-eyes" also refers to a sudden apprehension of similarity, seeing things in a new way, an epiphany, a eureka experience, or a shift in paradigms that changes how the world or that part of it is experienced henceforth. For Aristotle, then, metaphor, as an expression of *energia*, is linked with the creation of knowledge.

For Aristotle, "bringing-before-the-eyes" is germane to understanding the importance of metaphor. He emphasizes that by "bringing-before-the-eyes," he refers to "those things 'before the eyes' that signify activity" (1991, p. 248). He recognizes that "to say a good man is 'foursquare' is a metaphor . . . but it does not signify activity [*energia*]. But the phrase 'having his prime of life in full bloom' is *energia*" (p. 249). After discussing a series of examples from Homer, Aristotle concludes, "He [Homer] makes everything move and live, and *energia* is motion" (p. 249). Certainly for the nineteenth- and early twentieth-century physicists, the SSA brought the structure of the atom to life, something that they could not see but only detect the presence of with photographic plates and Geiger counters. Though it is still difficult to observe at the atomic level, today's experiments are largely instrumental, and the strings of string theory cannot be observed at all (Greene, 2003). In the nineteenth and early twentieth centuries, classical physics still reigned, so metaphors were all the more necessary to shape theory.

Aristotle views metaphor, then, as playing a role in the creation of knowledge. He also discusses analogy, the type of metaphor he finds most useful, with enthymeme, so he ties it to argument as well. Doing so more clearly includes metaphor as a part of Aristotle's *Rhetoric.* It seems, then, that according to Aristotle, metaphor can be persuasive as well, and it is persuasive because of *energia.* Since metaphor is persuasive because of its *energia*, then it follows that *energia* causes metaphor to be epistemological since knowledge is created to some degree by persuasion.

Ironically, the failure of Aristotle's science can be read as a rhetorical failure since he was not able to travel beyond that which was demonstrable, as McMullan (1976) has pointed out. In other words, Aristotle's science was not metaphorical in the sense that he did not create theory beyond observation. An object dropped, according to Aristotle, seeks its point of origin, which did not satisfy Newton, who sought theory to explain how an object, such as his famous apple, moves from zero velocity as it accelerates toward the ground. Newton then extrapolated this theory to explain the relationship between Earth and other celestial bodies.

Furthermore, to Aristotle's credit, his discussion of analogy can be found in his *Topics* (1952a), which can be read as more likely a part of his scientific epistemology because he privileged the syllogism in scientific discourse as a way of arriving at knowledge, rather than the enthymeme.

There is some evidence of Aristotle functioning as a technical communicator and using metaphor to describe a medical treatment of the day. Ironically, this example occurs when he discusses riddling. Riddles can be read as similar to metaphors in that they pivot upon what Beardsley (1962) refers to as "the metaphorical twist." On "good riddling," Aristotle (1991) tells us that "it is generally possible to derive appropriate metaphors; for metaphors are made like riddles; thus, clearly a metaphor from a good riddle is an apt transference of words" (pp. 224-225). First, it is interesting to note that Aristotle likens metaphors to riddles. What must appeal to him to make such a comparison is the type of *energia* created in what he would recognize as Beardsley's metaphorical twist. And of course, the ancients were not strangers to riddles: Oedipus' answer to the Sphinx's riddle allowed him to defeat it, so riddling has presence in ancient thought as demonstrating mental prowess, as perhaps a symbol of the results of arriving at an epiphany and experiencing new knowledge.

Aristotle's example of a riddle in this passage is no mere joke, though. He writes, "a man gluing bronze on another with fire" to describe the bleeding a physician might do "with a hot bronze cup that draws at blood as it cools." Aristotle adds, "the application of the cupping instrument is thus called gluing" (p. 224). Here, Aristotle is using a metaphor in the shape of a riddle to explain a medical process of his day. The reason for this particular riddle/metaphor is not known in terms of whether it exists for amusement or whether its purpose is to explain a medical process that some might find objectionable and belies common sense, just as in more recent times, many people have objected to organ transplants, pacemakers for heart patients, and today, to cloning. Of course, we now know that being bled really is not good for a patient, but at the time it was literally "cutting edge" medical science. Perhaps this example is evidence of Aristotle's use of metaphor to explain a medical process to the general public.

Some comparison might be made with another ancient rhetorician, such as the anonymous author of the *Rhetorica ad Herennium* (1990), since this work is the longest complete Latin rhetoric text, and it has the longest book on style (Bizzell & Herzberg, 1990). Such a comparison is important because the *Rhetorica ad Herennium* reflects the influence of Aristotle's *On Rhetoric* (1991) and represents the misinterpretation of metaphor manifested by the substitutionists to the extent that, as P. Ricoeur (1975) has claimed, the study of metaphor virtually died in the nineteenth century.

A cursory examination might suggest that as a handbook, the *Rhetorica ad Herennium* is superior to Aristotle's *On Rhetoric*. After all, it fulfills many contemporary notions of a handbook. For example, the *Rhetorica ad Herennium* is easier to read, one reason being that it is more clearly organized. Each paragraph begins with a topic sentence that is then illustrated with one or more examples. Aristotle's *On Rhetoric*, on the other hand, nearly requires a detective's scrutiny to follow his discussion of metaphor through chapters two, three, four, ten, and eleven of Book Three on "Delivery, Style, and Arrangement." However,

I would argue that in *On Rhetoric*, Aristotle succeeds in creating a philosophy of metaphor by casting it into a logical, enthymematic framework retaining a sense of "clarity, sweetness, and strangeness" (p. 223), while the *Rhetorica ad Herennium* functions on a more superficial level. Unfortunately, many rhetoricians who have followed Aristotle all the way to the present have been more similar to the author of the *Rhetorica ad Herennium* than to Aristotle. The respective portions on figurative language are areas where this dichotomy is especially apparent. Aristotle, on the one hand, praises metaphor for the concept of *energia* and moves metaphor into the realm of philosophy by indicating how it is related to epistemology. Conversely, the *Rhetorica ad Herennium* merely catalogs different types of metaphors.

Certainly the two texts share similarities in their approaches to figurative language. After all, the author of the *Rhetorica ad Herennium* wrote his treatise over 200 years after Aristotle's *On Rhetoric* (1991), and, despite the Roman disdain for "greeklings," probably read Aristotle's work, but the two works differ in complexity. For example, the definitions of metaphor in *Rhetorica ad Herennium* and in Aristotle's *Rhetoric* are roughly similar. The *Rhetorica ad Herennium* defines metaphor as "when a word applying to one thing is transferred to another, because the similarity seems to justify this transference" before noting how metaphor can render "a vivid mental picture" (p. 278). This idea of a metaphor creating a visual image is, to this day, according to Ricoeur (1975), leftover from sorcery, since meaning is found in discourse, not in semantics.

Though Aristotle defines metaphor as "the transference being either from genus to species or from species to genus or from species to species or on the grounds of analogy" (1952b, p. 693), his basis of comparison is much more complex. In terms of invoking a comparative element to define metaphor, these two definitions are similar; but while the *Rhetorica ad Herennium* suggests that metaphor occurs when "the similarity *seems to justify* [my italics] this transference" (p. 278), Aristotle specifies for metaphor a more specific model (A:B) that can be applied to and expanded upon in his discussion of analogy. For Aristotle, the importance of metaphor lies in its creation of *energia* that leads to new knowledge; the author of *Rhetorica ad Herennium* merely observes the presence of metaphor but seems unsure of its value, other than for creating "a vivid mental picture," "for the sake of avoiding obscenity," for "magnifying" or "minifying," or "embellishing" (p. 278). Furthermore, the author cites Cicero, Aristotle, Longinus, and Quintillian, who warn that metaphor should be "restrained" (p. 278). The author does not seem to value metaphor as a philosophical touchstone or to value it for any philosophical importance.

Both authors extend their discussion of metaphor to a more complex figure of thought. In the same chapter, the author of *Rhetorica ad Herennium* follows his discussion of metaphor with one of allegory that is defined as "a manner of speech denoting one thing by the letter of the words, but another by their meaning . . . and operates through comparison when a number of metaphors originating in a similarity in the mode of expression are set together" (p. 278). As an example, he poses the question, "For when dogs act the part of wolves, to what guardian, pray, are we going to entrust our herds of cattle?" (p. 278). Such an example seems only slightly more akin to the current notion of metaphor than to even the medieval concept of allegory as the interpretation of the symbolic significance of a work.

The *Rhetorica ad Herennium* also suggests that allegory may be useful as comparison or contrast for argumentative purposes, but nowhere in this discussion is there an attempt to draw metaphor into a meaningful philosophical unity with the rest of the work. Analogy itself is discussed as a "Figure of Thought," but its treatment is briefer than that allotted to allegory. Analogy really seems to be little more than comparison, especially with "Do not, Saturninus, rely too much on the popular mob—unavenged lie the Gracchi" as an example (p. 291). Aristotle, on the other hand, by interweaving his discussion of analogy through five chapters, integrates it more successfully into his rhetoric.

From this comparison, the *Rhetorica ad Herennium* lacks the depth of discussion apparent in Aristotle's *Rhetoric*. Unlike Aristotle, the anonymous author does not arrive at metaphor as an epistemological tool. Unfortunately, the *Rhetorica ad Herennium* seems more similar to current handbooks in how it divides and classifies metaphor. To illustrate this point, consider a recent edition of the *Harbrace College Handbook* (Hodges, 1998) that confines discussion of figurative language to brief definitions of terms, with examples of metaphor and simile. Other types of metaphor such as personification, paradox, overstatement and understatement are alluded to but not explained. Aristotle's approach to metaphor delves deeper. The substitutionist approach apparent at the time of the ancient Romans influenced contemporary use of metaphor in the sciences, where it is viewed with suspicion as a rhetorical ornament. This suspicion is now deeply embedded in Western culture. Aristotle is not to blame for this misinterpretation. Many rhetoricians such as Nietzsche (1990), Richards (1936), Ricoeur (1975), and Perelman and Olbrechts-Tyteca (1969) have proclaimed all language metaphoric, and certainly the way Aristotle elevated metaphor represents a value beyond the cataloging of types of metaphors.

Metaphor is at the root of created language, and we can see another example in Aristotle's work. Hyphenated words are another aspect of metaphor important to the scientist because they deal with the creation of words to name the unnamed, what Perelman and Olbrechts-Tyteca note as the fusing of the theme and phoros spheres (the equivalent to Richards' tenor and vehicle) and what Holton (1986) refers to as science's "fast metabolism" that causes it to grow "so much faster than other fields of thought and action" (p. 239). On this note, Perelman and Olbrechts-Tyteca have also claimed for metaphor that "The [ancient] masters of rhetoric saw in metaphor a means of overpowering the poverty of language" (p. 404). They have also named compound words as metaphoric, which is important to consider with the naming of objects and ideas, an arena where words and knowledge are often created.

The importance of naming is explained in *Poetics* when Aristotle (1952b) observes, "It may be that some of the terms thus related have no special name of their own, but for all that they will be metaphorically described in just the same way" (p. 693). Hence, metaphor may work to name the otherwise unknown. A hyphenated modifier defines by filling a semantic void. There is no one word that adequately describes either "biological-waste management system" or "biological waste-management system" (Alred, Brusaw, & Oliu, 2000, p. 256), so the shifting hyphens are somewhat metaphoric. Aristotle continues his discussion of this matter with an example:

> Thus to cast forth seed-corn is called 'sowing'; but to cast forth its flame, as said of the sun has no special name. This *nameless act* [my italics] (B), however, stands in just the same relation to its object, sunlight (A), as sowing (D) to the seed-corn (C). Hence the expression in the poet, 'sowing around a god-created flame' (1952b, p. 693).

The flame is not named a "god" flame, it is not simply a "created" flame, but a "god-created flame." It could be argued that "god flame" is sufficient, but "god-created flame" suggests a particular act by a god rather than flame as characteristic in itself of a god. This example is better in some ways than examining the nuances of the "biological-waste management system" or the "biological waste-management system" because it utilizes the verb "created." Such a verb more clearly serves as an *energia* that "brings-before-the-eyes" by its virtue as a verb, especially one that more clearly suggests an act. Though the hyphenation in this example is an English construction, it is in such a nameless act that language and knowledge can be created.

From this examination, hyphenated and compound words appear to have *energia*. I have noted the motion resulting from simply shifting the hyphen and how that motion shifts meaning. Such a distinction is important because it parallels the nameless act set into motion that creates a metaphor. Such a distinction is important to understanding scientific and technical communication and all thought as metaphorical. Moving from a noun to a verb suggests the motion of an epistemological metaphor.

I have mentioned that the misinterpretation of Aristotle continues today. Let us now consider some twentieth-century rhetoricians in whose work the substitution approach can be discerned.

TWENTIETH-CENTURY SUBSTITUTIONISTS

I. A. Richards' (1936) approach to metaphor is shaped by Aristotelian interpretations exemplified by the *Rhetorica ad Herennium*. On the history of rhetoric, Richards critiques the field as having been limited in the eighteenth century "to certain modes only," though he does not specify those modes. The nineteenth century emphasized other modes, and his half of the twentieth century is "recovering from these two specializations," an effort he thought of as "a way of reformulating the Classic-Romantic antithesis." "Such a recovery might proceed best," he poses, by "a better developed theory of metaphor than is yet available" (p. 94).

First, it is worth noting the extent that, like Aristotle, Richards praises metaphor. He states that "thought is metaphoric" (1936, p. 94), "we cannot get through three sentences of ordinary fluid discourse without it [metaphor]," and "the pretense to do without metaphor is never more than a bluff waiting to be called" (p. 92). Furthermore, he extends metaphor to the sciences:

> Even in the rigid language of the settled sciences we do not eliminate or prevent it without great difficulty. In the semi-technicalised subjects, in aesthetics, politics . . . our constant chief difficulty is to discover how we are using it and how our supposedly fixed words are shifting their senses (p. 92).

Though Richards indicates an interest in examining metaphoric usage, this passage provides further evidence as to why he should be read as a substitutionist, since he seems to believe, despite his declaration that "all language is metaphoric," that there is a usage of language in the "settled sciences" that is not metaphoric. His call for greater attention to the workings of metaphor never leads to a theory beyond substitution. Though he states that "thought is metaphoric," he favors the "transfer [of] more of our skill into discussable science" (p. 94). Unfortunately, his articulation of the scientific approach suggests that there are uses for words that are not metaphoric, and he supports this view when he claims a word can be "simultaneously literal and metaphoric," which suggests a literal fixed meaning and a separate metaphorical meaning. Such an assertion contradicts his idea of all language as metaphorical. On this point, Johnson-Sheehan (1999) more recently pointed rhetoricians in the direction of metaphor as a hermeneutic rather than a linguistic construction.

Richards does add to the development of metaphor theory by naming parts of the metaphor the "tenor" and the "vehicle." Though Aristotle began metaphorical analysis with his model of the A:B structure, Richards calls more attention to the parts by naming them and supplies "vehicle," which, though a noun, suggests *energia*. The "tenor" is that which is to be described; the "vehicle" is that which enables the metaphor. With a metaphor such as, "An atom is a miniature solar system," the "atom" is the "tenor," the unknown entity to be defined metaphorically. The "vehicle" is that which makes the metaphor possible, the "solar system," in this case. With a more pedestrian metaphor, such as the oft-cited "Man is a wolf," the audience knows both entities, but the application informs the audience of the writer's insight into humanity.

Another way in which Richards adds to the discussion of metaphor is through what Karl Popper (1972) (who is discussed later in this chapter as contributing to metaphor as an epistemology) would call falsification, the type of dialectic the scientist enters through testing and verification of findings. Richards notes that

> Once we begin 'to examine attentively' interactions which do not work through *resemblances* between tenor and vehicle, but depend upon other relations between them, including *disparities*, some of our most prevalent, over-simple, ruling assumptions about metaphor as comparisons are soon exposed [Richards' italics] (1936, pp. 107-108).

The "disparities" that Richards mentions are the aspects that are not comparative, but present in the metaphor; though people share some characteristics with wolves, we are not wolves, and our moral and ethical imperatives cause us to strive to be otherwise. With the scientific metaphor, the differences, or the ways in which the metaphor might be falsified, might also be points with epistemological significance. An atom is a miniature solar system in that the nucleus, like the sun, is its center, and the subatomic particles move around it. Some of the particles can even be said to be in predictable "orbits," but not all. The problem is that the subatomic particles do not radiate energy, so according to classical physics, they should fall into the nucleus. To explain this conundrum, quantum mechanics theorizes that the subatomic particles jump from orbit to orbit. Though it has weaknesses as a model, the SSA can contribute to the creation of knowledge. And it is worth noting that the "orbit" metaphor remains in place today, even with string theory, which hypothesizes what makes up atoms.

Though Richards uses the word "interactions," he does not follow through with it to a theory of metaphor. Instead, Richards would relate metaphor to the creation of language and the creation of knowledge:

> The processes of metaphor in language, the exchanges between the meanings of words, which we study in explicit verbal metaphors, are super-imposed upon a perceived world which is in itself a product of earlier or unwitting metaphor . . . That is why, if we take the theory of metaphor further than the 18th Century took it, we must have some general theorem of meaning (1936, pp. 108-109).

Richards, then, would take metaphor a step further than Aristotle in terms of its philosophical import when he expresses a desire to create a theory to describe the workings of metaphor. What is lacking is an attempt to construct a coherent theory beyond the idea of tenor and vehicle.

Perelman and Olbrechts-Tyteca (1969) can be read as substitution theorists as well in their study of metaphor. Like Richards, they break metaphor down and name its parts, using "theme" for Richards' "tenor" and "phoros" for Richards' "vehicle." "Phoros" is "coined from the Greek *phoros*, or 'bearing,' which is found in *metaphora*" (1969, p. 373). They also recognize the problems of the substitution approach: "Because of the tendency of rhetoricians to restrict their study to problems of style and expression, rhetorical figures increasingly came to be regarded as mere ornaments that made the style artificial and ornate" (p. 167). Richards, according to Perelman and Olbrechts-Tyteca, was not satisfied with a substitution or comparison theory of metaphor because it is "misleading and inadequate. . . . To him metaphor is much more a matter of interaction than of substitution, a technique of research as much as one of embellishment" (1969, p. 399). Their recognition of Richards' dilemma suggests that they will create a theory of metaphor. However, their approach to metaphor is based on syntax and does not move beyond substitution theory, despite what appear to be better intentions.

That Perelman and Olbrechts-Tyteca intended their work to be read as Aristotelian is evident from the way they parallel analogy with enthymeme:

> The richest and most significant metaphors are not, however, like those of Plotinus or Ronsard, which arise out of the expression of an analogy, but those that are from the outset presented as metaphors generally by coupling the superior terms of the theme and the phoros (A and C) and leaving unexpressed the inferior terms (B and D) (1969, pp. 400-401).

Perelman and Olbrechts-Tyteca warn, "If the argumentative role of figures is disregarded, their study will soon seem to be a useless pastime, a search for strange names for rather farfetched turns of speech" (1969, p. 167). Though such criticism of what became the traditional approach to metaphor is pertinent, and pointing toward metaphor as a tool for argument is a step that could be theoretically generative, Perelman and Olbrechts-Tyteca do not travel quite far enough with metaphor.

They contribute to an enriched sense of metaphor by examining the many ways that it can be posed. Ironically, the regimented methodological discipline attributed to science is, fortunately, not practiced in many scientific analogies, which increases the significance of Perelman and Olbrechts-Tyteca's contribution. For analogy in the sciences, they note "an asymmetrical relation between theme and phoros, arising from the position they occupy in the spheres" (p. 373). They recognize this inconsistency as what I refer to as "elasticity," by which I mean the many ways the A:B::C:D structure can be varied. For example, Perelman and Olbrechts-Tyteca observe, "when we say that every analogy involves a relation among four terms, we are, of course, giving a schematized picture of things. In fact, each term may correspond to a complex situation, and such a situation is precisely what makes a rich analogy" (p. 375). Such an explanation is helpful to keep in mind when the complexity of some scientific analogies is considered.

Scientific analogies may go quite far beyond the A:B::C:D structure. Consider the elasticity of this analogy:

> An electrical circuit's design parallels a system of water pipes. The water flows along the pipes like an electrical current flows along wires. A reservoir supplies water like a power source supplies electricity. Dams retain water like resistors retain electricity. Stored water in a reservoir is potential energy, just as voltage in a circuit is potential energy (Gentner & Gentner, 1983).

Such a rich analogy extends far over the A:B::C:D paradigm into mapping a much more complex structure:

A:B::C:D
C:D::E:F
G:C::H:E
H:C::I:E
C:J::K:J

The arrangement weaves a complicated tapestry associated more with literature than science, but literary prose analogies are less frequently extended in this fashion and are intended more often to create an ambience in the moment of the text or as motif contributing to theme than as an explanatory vehicle. Perelman and Olbrechts-Tyteca comment, "Rich analogies can be drawn with the aid of double hierarchies, as these are characterized by complex relations both horizontal and vertical—the former based on the structure of reality, the latter exhibiting a hierarchic progression" (1969, p. 377). When this aspect of Perelman and Olbrechts-Tyteca's work is considered, they seem to be moving in the direction of the interactive aspect of metaphor, but they focus more on the structure of the sentence than on the verb, so ironically, their work is more Aristotelian than they might intend. One could argue that the structure of the sentence is a move toward interaction, but the adherence to discussion of the A:B::C:D structure is very Aristotelian. Unfortunately, Aristotle's concern with *energia* is not apparent, beyond the workings of theme and phoros.

Perelman and Olbrechts-Tyteca (1969) have written more specifically on metaphor in science. They suggest the idea of an antimodel as negative reinforcement, and they note that the use of analogy in philosophy has shifted. Perelman himself (1982) has cited Plato and St. Thomas for basing arguments on analogies, but for empiricists, analogy "is limited to affirming a weak resemblance and is useful for formulating hypotheses, but must be eliminated in the formulations of the results of scientific research" (p. 114). A type of logical fallacy often cited in grammar handbooks is called a "false analogy," and writers are warned to avoid basing an argument on an analogy because eventually it breaks down. An atom, after all, is not a miniature solar system. However, the value of the negative analogy should not be discounted since it too may lead to knowledge, to which I have alluded and explore further when I move to the epistemology of metaphor.

Though analogy may be a tool of argument, it should not be expected to correspond to reality, as the scientist expects. Perelman (1982) comments,

> In certain cases, after an analogy has allowed a scientist to orient his investigations, which in turn have given experimental results according to which he will structure them independently of the *phoros*, he can abandon the analogy, as does the contractor who takes down the scaffolding after the building is finished. Thus, after the analogy established between electric and hydraulic current gave direction to the first experiments in the field, further experimentation could finally develop in an independent way.
>
> In other cases, the analogy will be surpassed, theme and phoros both being reduced to a more general law. But in fields where recourse to empirical methods is impossible, analogy cannot be dispensed with, and the argument that is used will be employed mainly to support it and show its adequacy (p. 115).

According to Perelman and Olbrechts-Tyteca (1969), then, analogy serves a very utilitarian function until it reaches the point where empirical observation causes it to be no longer meaningful. Where empirical observation fails, analogy can succeed with giving shape to models that would otherwise be ineffable.

Such was the value of the SSA (Solar System Analogy) until Bohr dispensed with it in favor of a quantitative model that is still, nonetheless, a model. Kuhn (1993) has pointed out that the SSA was never meant to be an exact model. The point was to determine in what ways the SSA described the structure of the atom and in what ways it did not. Kuhn explains why the SSA still has value:

> Furthermore, even when that process of exploring potential similarities had gone as far as it could (it has never been completed), the model remained essential to the theory. Without its aid, one cannot even today write down Schrodinger's equation for a complex atom or molecule, for it is to the model, not directly to nature, that the various terms in that equation refer (p. 538).

Perelman and Olbrechts-Tyteca's observations seem very cut and dried and obvious, but at what point does theory become empirical? If it is completely empirical, then is it theory? Theory accounts for data, but at some point, especially on the frontiers of science, the scientist must again make a Kierkegaardian leap of faith. In this leap, we find metaphor.

More recently, Perelman (1989) has recognized that the scientist uses metaphor to generate theory. Specifically, analogies "play an essentially heuristic role as instruments of invention; they give the researcher hypotheses to organize his investigations. Their fecundity, the new perspective that they open to the researcher, gives them their importance" (p. 82). On the other hand, he posits that

> Eventually, however, they [metaphors] must be put aside; the acquired results must be formulated in technical language, whose terms must be gotten from the specific theories of the investigated field. Ultimately, analogy will be replaced by a model, a schema or a general law which encompasses theme and phoros (p. 82).

Furthermore, Perelman (1989) claims, "mathematical procedure is preferential allurement" (p. 82), and he and Olbrechts-Tyteca have argued that "For centuries many good minds have found in the artificial language of mathematicians an ideal of clarity and univocity that natural languages, with their lesser development, should strive to imitate" (1969, p. 130), without recognizing that mathematics, as an artificial language itself, is nothing but another type of metaphor. In addition, by focusing to such a great degree on the way metaphor is ultimately discarded, they are functioning as literal substitutionists who have misread Aristotle's *Rhetoric* and contribute to negative attitudes toward metaphor. If rhetoricians do not value metaphor, then why should scientists?

The process of becoming a dead metaphor as a way that language invents itself is what Perelman and Olbrechts-Tyteca refer to as the "outstripping" of the metaphor. However, such a change does not carry with it the usual negative connotations. In this case, that which is destroyed emerges phoenix-like as a law. Perelman and Olbrechts-Tyteca comment, "If the analogy is a fruitful one, theme and phoros are transformed into examples or illustrations of a more general law, and by their relation to this law there is a unification of the theme and the phoros" (1969, p. 396), which points to metaphor's generative quality as more in keeping with the true spirit of the Aristotelian metaphor.

What bothers Perelman are metaphor's gray areas: "Every analogy highlights certain relationships and leaves others in shadows. With good reason Max Black [1962] has emphasized that describing a battle with terms borrowed from checkers disregards all the horrors of war" (1982, p. 119). It is ironic that Perelman would cite Black, because he is an interactionist, a theorist who is more interested in how a metaphor works (or does not work). Unlike Black, Perelman and Olbrechts-Tyteca (1969), and Richards (1936) read metaphor as reaching a point of failure rather than an epistemological base of thought and language.

I have established, then, reasons for discussing rhetoric (and metaphor) concurrently with philosophy. The reasoning lies in the seventeenth-century validation of probability. Though Bacon (1952) is credited with faulting metaphor, he emerges as one who would gladly use it as a rhetorical tool, especially when the sciences would be communicated to a general audience. Aristotle analyzed and presented metaphor as worthy of philosophy. His influence on Richards (1936) and Perelman and Olbrechts-Tyteca (1969) has also been noted as causing them to approach the study of metaphor as substitutionists. However, Nietzsche has also influenced these rhetoricians, so I must next turn to his work to approach the postmodern perspective of metaphor studies.

NIETZSCHE AND POST-MODERN METAPHOR

Nietzsche (1989, 1990) can be read as a transitional figure in the study of metaphor. He was the first to claim all language as metaphoric: "what is usually called language is actually all figuration" (1989, p. 25), and, "with respect to their meanings, all words are tropes" (1989, p. 23). His work certainly influenced twentieth-century rhetoricians and is reflected in many of their writings, especially Richards (1936), Perelman and Olbrechts-Tyteca (1969) and Weaver (1990).

"What is a word?" Nietzsche asks, "It is the copy in sound of nerve stimulus" (1990, p. 890), so Nietzsche as well would subtract the boundaries of definition from notions of a word and reduce it to not only the most primal of utterances but even further to tremors along a strand of nerve. According to Nietzsche, what we think we know about words is connected to the idea that "every word instantly becomes a concept insofar as it is not supposed to serve as a reminder of the unique and entirely individual original experience to which it owes its origin" (1990 p. 891). As an example, he poses the idea of a leaf. All leaves are different, but the concept of a leaf allows us to look at a leaf, recognize it, and name it, to look at an oak leaf and a pine needle and understand their similarities and to ignore the differences. Being able to do so is related to the concept of metaphor.

Richards (1936), Perelman and Olbrechts-Tyteca (1969), and Weaver (1990) share a greater similarity with their approaches to metaphor, and in that sense, they depart from traditional Aristotelian rhetoric and align more neatly with Nietzsche. Weaver, the most Aristotelian in his approach to persuasion and dialectic, suggests that at the point where a rhetorician has brought his audience to the "truth," "there is no way to move them except through the operation of analogy" (1990, p. 1061). Metaphor, according to Weaver, not only is persuasive as a rhetorical mechanism, but furthermore, it provides the impetus toward "a cure of souls . . . toward an ideal good" (p. 1062).

If Nietzsche had been cognizant of what is now called existentialism (and considered himself an existentialist), he would praise Weaver for such an intuitive leap of faith. Weaver (1990) recognizes that "when the disputed terms have been established, we are at the limit of dialectic" (p. 1061). For Weaver, the point of dialectic is to arrive at an ultimate definition for terms germane to an argument, but the final step may be found in metaphor.

Because Perelman and Olbrechts-Tyteca's (1969) distrust of metaphor, especially for science, has been noted, it is uncertain to what degree Nietzsche has influenced them. As a matter of fact, they seem to be engaging Nietzsche in an oblique dialectic. For example, in *The New Rhetoric,* Nietzsche is referred to only a few times, and in two of the instances, he is not discussed by Perelman and Olbrechts-Tyteca but only mentioned in quotations from other authors whose work the authors cite as examples. When Nietzsche is mentioned, it is as a "popular philosopher," his work unlike the "contemporary philosophies which all presuppose a thorough knowledge of the history of philosophy" (p.100). This slight is but a prelude to their discussion of dead metaphor.

When Perelman and Olbrechts-Tyteca discuss dead metaphor, they are considering what Nietzsche would call "truths" that "are illusions which we have forgotten are illusions; they are metaphors that have become worn out and have been drained of sensuous force, coins which have lost their embossing and are now considered as metal and no longer as coins" (1990, p. 891). Perelman and Olbrechts-Tyteca note that "outstripping an analogy has the effect of making it appear as the result of a discovery, as an observation of what is, rather than as the product of an original effort at structuration" (p. 397). In other words, when an analogy becomes a dead metaphor, the fact that it was once metaphor has been lost. Then, Perelman and Olbrechts-Tyteca point out that "In some cases the problem is reversed" (p. 397). The term is turned back into a metaphor, which is what Nietzsche has done when he claims that all language is metaphoric. The SSA, for example, remains in a metaphoric state since it is inaccurate, but it persists as a teaching tool.

Perelman and Olbrechts-Tyteca include a discussion of monism. They continue with, "There are philosophies which consider analogy as the result of differentiation within a unitary whole; this is true of monistic philosophies which refuse to allow any distinction between fields" (p. 397). Actually, they seem closer to monism than Nietzsche, since they have not traveled as far as Nietzsche from considering analogy as a rhetorical entity more powerful than a tool to be used and discarded. However, their approach does not account for an epistemology of metaphor, a field Nietzsche foraged as he lent to metaphor the ability to create language, which is related to metaphor's ability to create knowledge. Nietzsche waxes epistemological when he speaks of the role of the scientists in relation to language and how it "works on the construction of concepts . . . just as the bee simultaneously constructs cells and fills them with honey, so science works on this great columbarium of concepts, the graveyard of perceptions" (1990, p. 894). To create knowledge, then, science must work from what language has created through metaphor: "The scientific investigator builds his hut right next to the tower of science so that he will be able to work on it and to find shelter for himself beneath those bulwarks which presently exist" (Nietzsche, 1990, p. 894), so the scientist is directly involved in the construction of concepts through involvement in the construction of language, and therefore, of knowledge.

Perelman and Olbrechts-Tyteca (1969) counter that Nietzsche (1990) could not possibly be serious since he is not examining fully the many ways in which words are created and used. To support this idea, the authors conclude, "however, these philosophical considerations with respect to the status of analogy do not, in practice, disturb the normal possibilities of using analogy and its tendency to be outstripped" (p. 397). The difference for Nietzsche lies in the meaning of what

Perelman and Olbrechts-Tyteca refer to as outstripping. For them, outstripping refers to a metaphor's mortality, but for Nietzsche, this "outstripping" is generative, and Perelman and Olbrechts-Tyteca contradict themselves somewhat when they discuss how a metaphor may be resurrected. When they continue with, "the analogy then merely makes explicit that which was included in the undifferentiated whole that preceded it" (p. 397), they are saying that the analogy is supplementary to meaning.

From this examination, it is evident that Perelman and Olbrechts-Tyteca would subsume that which Nietzsche would elevate. That which they would ignore, however, is too volatile to be considered permanently dead, by their own admission.

I. A. Richards (1936) is another rhetorician under Nietzsche's sway. As I have noted, when Richards asserts that "thought is metaphoric" (p. 94) "we cannot get through three sentences of ordinary discourse without it [metaphor]" (p. 92), and "the pretense to do without metaphor is never more than a bluff waiting to be called" (1936, p. 92), he is echoing Nietzsche, who said "what is usually called language is actually all figuration" (1989, p. 25), and that language itself is metaphoric (p. 24). While such an assertion is bold, Nietzsche takes it step further and relates it to truth, which is but "a moveable host of metaphors . . . a sum of human relations . . . poetically and rhetorically intensified" (1990, p. 891). Truth is but a forgetting of the fact that what separates us from animals is based upon our cognition of metaphor and therefore upon a lie, according to Nietzsche. The result is that there are two kinds of people, those whom Nietzsche calls the rational and those whom he calls the intuitive. The rational are ruled by reason that abides by law expressed and defined through language, so "as a *rational* being, he now places his behavior under the control of abstractions" (1990, p. 892). The intuitive, however, though "he suffers more frequently since he does not know how to learn from experience and keeps falling over and over again into the same ditch" (1990, p. 896), becomes an innovative artist,

> who speaks only in forbidden metaphors and in unheard of combinations of concepts. The intuitive one does so because by shattering and mocking the old conceptual barriers, he may at least correspond creatively to the impression of the powerful present intuition (p. 895).

These new expressions of metaphor strive, according to Nietzsche, to push against the lie accepted as meaning, to vainly search for meaning where none can be found because it can be expressed only in words, which are only "the copy in sound of the nerve stimulus," not the thing itself (p. 890). Richards, then, more neatly aligns with Nietzsche. However, Richards does not take the final step. Though he claims all thought as metaphoric, he seems to believe that in the sciences, metaphor can somehow be expunged.

Nietzsche's impact on a rhetoric of metaphor is profound, considering the small amount of work he completed specifically focused on metaphor. While Weaver (1990) and Perelman and Olbrechts-Tyteca (1969) differ in the degree to which they would endorse Nietzsche's notion of definition, all of these twentieth-century rhetoricians engage, to some degree, Nietzsche's idea of metaphor, restoring it to respectability as a rhetorical tool, and not simply a literary device, though it has not been recognized completely as such. Recognizing the power of metaphor enriches language, especially in science and technology where, ironically, Perelman and Olbrechts-Tyteca's universal audience demands more from cross-discipline communication.

Not delving any further than substitution theory is symptomatic of a deeper problem in rhetorical studies because substitution theory is but a framework upon which to stretch the prose in an interesting and ornamental way. Perelman and Olbrechts-Tyteca noted that this tendency caused rhetoricians to focus on the many different types of tropes, of which metaphor is but one. The focus then becomes dividing and classifying rather than studying how and why metaphor works. When a metaphor works well in the sciences, it contains P-fertility (proven fertility), as McMullan (1976) noted, and it becomes useful to scientific theory; only when it does not work, according to Perelman and Olbrechts-Tyteca, should it be regarded as mere ornamentation. When rhetoric is taught as the art of delivering a speech well or writing a five-paragraph essay with all of the commas in the right place, it is analogous to a theory of metaphor that divides and classifies rather than examining how and why metaphor might work, especially when its relationship to the creation of knowledge is considered. Knowing how and why metaphors work is important to scientists since science is so rife with metaphors. To what extent might metaphor aid the scientist in creating science? To what extent might a subconsciously used metaphor misdirect science? Study of metaphor can answer these questions for a budding scientist.

Ricoeur's (1975) perspective on metaphor, especially as his work relates to Aristotle, makes for a fitting conclusion to this discussion of substitutionists. Ricoeur argues that substitution as the point of metaphor is responsible for casting rhetoric in such a fashion that it died during the nineteenth century. Ricoeur focuses on the root of the problem as being as basic as the part of speech assigned to metaphor. So long as metaphor remains a noun, it is susceptible to being cast as a substitution or comparison. Such was not Aristotle's intent, according to Ricoeur, who emphasizes that Aristotle's theory of metaphor creates bringing-before-the-eyes, which is part of the epistemology of metaphor since such movement is also related to how metaphor can influence audiences by shifting their perspective. Ricoeur argues that a literal translation of what Aristotle sees as the strength of metaphor, that its master can perceive and express the similarity and dissimilarity, would have metaphor as the clause's verb, "metaphorize." Recognizing it as a verb realizes the motion Aristotle intended for this term and prevents it from becoming a noun that suggests substitution, or items that are divided and classified. Such an approach causes metaphor to become mechanical rather than philosophical. Aristotle is not blameless, however, according to Ricoeur, over the status of rhetoric, since it was Aristotle who emphasized argument and composition in his *Rhetoric*, lumping metaphor into a general category of style.

Nietzsche (1989, 1990), then, is responsible for postmodern attitudes toward metaphor by placing it at the seat of language. Twentieth-century rhetoricians such as Richards (1936), Weaver (1990), and Perelman and Olbrechts-Tyteca (1969) have struggled with their desire to fit into what they perceive as the Aristotelian tradition that casts its long shadow over rhetoric. On the other hand, they have rebelled against such theory perhaps because they sense the interactive quality. Unfortunately, with the substitutionists, the interactive quality is not yet realized in meaningful theory. Ricoeur (1975) maintains that what seems like the traditional Aristotelian treatment of metaphor is actually a misreading that has miscast the role of metaphor. I next examine the tensionists, who serve as an introduction to the interactionists, and explore the interactive quality Richards yearned for but did not realize.

THE TENSIONISTS: AN INTRODUCTION
TO INTERACTION

The tensionists further broke metaphor down to analyze how its constitutive parts worked rather than simply being satisfied with naming the parts. I begin by considering M. C. Beardsley's (1962) "metaphorical twist."

What interests Beardsley is how metaphor actually works. He is interested in metaphor at a syntactic as well as semantic level. On semantics, he differs from Richards (1936) and Perelman and Olbrechts-Tyteca (1969). First, the syntactic level can be characterized by what Beardsley refers to as a "metaphorical twist" that occurs in the predicate. To illustrate this idea, the SSA can be posed more clearly as an analogy: "As the planets orbit the sun, the subatomic particles orbit the nucleus." For Beardsley, his metaphorical twist is an event and a meaning, which can be interpreted as moving toward an epistemology of metaphor. This new meaning makes the metaphor significant in the sense that McMullan (1976) characterizes as fertility. Furthermore, Beardsley (1962) sees metaphor as moving away from a central, accepted meaning to a meaning that becomes marginal and opposes the logical or accepted meaning. However, marginality does not suggest a lack of significance; to the contrary, it alludes instead to Aristotle's "strangeness," or novelty.

Beardsley (1962) is not alone here. The metaphorical twist represents what de Man (1979), echoing Nietzsche (1989, 1990), would call language as a lie. Metaphor is a lie because even with the fertile SSA, it is not literally true. Quantum mechanics, which can certainly be read as simply another metaphor, was necessary to represent what late nineteenth- and early twentieth-century physicists were not willing to allow with the SSA: to permit the subatomic particles to jump from one orbit to another within the metaphorical scope of the SSA. However, this analogy lives on in most secondary school science textbooks, as presented in chapter four.

Exploring metaphor at a deeper level, Beardsely (1962) sees it as standing at the frontier of language where meaning is created. He would agree with I. A. Richards' (1936) maxim that "All language is metaphoric," but would disagree with Richards' idea of the tenor and vehicle as useful ways of identifying the parts of the metaphor. Such an approach, according to Beardsley, is simply another form of substitution theory, so Beardsley would posit Perelman and Olbrechts-Tyteca's (1969) theme and phoros as substitution theory as well. Richards' neo-Aristotelianism does not satisfy Beardsley, who reads such an approach as one that focuses on metaphor as a shuffling of objects. Instead, Beardsley is more interested in the type of change the metaphor undergoes in terms of the tension created by the subject and predicate relationship.

Beardsley (1962) should not be read as entirely rejecting Aristotelianism. In a sense, Beardsley, too, is an Aristotelian because what interests him is the movement, the *energia* of the metaphorical moment. The connotation of such a moment creates a context for experiencing the metaphor that becomes more intuitive, and in that sense Nietzschean, than substitution theory allows. What Beardsley clarifies is what appears contradictory in Aristotle's approach to metaphor in that Aristotle marvels at metaphor, calling it a mark of genius and something that cannot be taught, and also claiming that it aids teaching because it most brings about learning, though Aristotle's A:B::C:D approach also seems to be an attempt to teach analogy as a mechanical entity. To what extent, though, is Beardsley's approach applicable to analogy, a clearer, though more complicated, case of substitution? Is the subject-predicate relationship better suited to metaphor than analogy? For example, with "As the planets orbit the sun, the subatomic particles orbit the atom's nucleus," the elements that correspond to A:B::C:D are definitely treated as objects that substitute one for another. Is there a metaphorical twist? Between A and B, with A corresponding to the planets and B to the Sun, it is reasonable to suppose there is no metaphorical twist. Aristotle would like the one-to-one correspondence of the subject to the predicate. This statement can be further demonstrated through mathematical models and through empirical observation. However, Beardsley would read two twists into this analogy. On the clause level, there is a twist between the subatomic particles, C, and the atom's nucleus, D. The relationship between the nucleus and the subatomic particles is still, though predictable, not empirically known. On the sentence level, there is another twist, one of Aristotelian *energia*, that occurs as an understanding of atomic structure is brought before the mind of the reader by the comparison of the solar system's structure to the atom.

Beardsley (1962) recommends studying metaphors more carefully to learn how they work. Scientific metaphors are especially appropriate because they are less likely to be dressed in additional ornamentation and are intended to explain data with a theory. They also "live long lives," as Johnson-Sheehan has pointed out (1998, p. 177). Indeed, the SSA has long been valuable, beginning with its U-fertility (unknown fertility) in 1865, and it continued to be valuable through Bohr's work around 1913. It is still around today, in various places, such as the CD-ROM version of *World Book Encyclopedia* and most secondary-school textbooks.

Beardsley would probably agree that his theory of the metaphorical twist has implications for scientific metaphors. What is most pertinent is the way that new words, or new definitions for ones currently in use, are created meta-phorically. While there is not much application at the level of these examples, Beardsley's ideas become more apparent if the development of a word such as "cell," from a hut attached to a monastery to a name for a microscopic organism, is considered. Ironically, though the metaphorical twist seems to be a semantic one, its metaphorical significance arrives at the syntactic level of usage, which means it must be cast where the twist occurs.

Ricoeur (1975) would agree with Beardsley (1962) on this point. According to Ricoeur, one function of metaphor is that it "fills a semantic void" (p. 17). Aristotle supports this idea as well when he writes of the sun flinging its rays as a nameless act. Ricoeur has argued Aristotle's nameless act is named by metaphor.

D. Berggren (1962/1963) has also addressed the idea of metaphoric tension. In his essays "The Use and Abuse of Metaphor, I & II," he begins with literature and the idea of tension between the metaphor and the idea of stereoscopic vision, which he defines as, "the ability to entertain two different points of view at the same time," an idea he attributes to W. B. Stanford as a necessary way to read in general (p. 243).

On myth, Berggren (1962/1963) writes, "The ultimate conclusion to be defended is that while creative thought in all of these areas is inescapably metaphorical in the sense to be defined, the tendency to abuse metaphor by transforming it into myth is no less prevalent" (p. 238). This transformation belies the way Berggren would divide and classify metaphor into the categories of the physical (models), schematic (what Black (1962) refers to as intuitive, such as Einstein's people at the train station), or formal (analogies drawn between nature and a machine). Berggren (1962/1963) concludes, however, that

> It is precisely this transformation of both referents, moreover, interacting with their normal meanings, which makes it ultimately impossible to reduce completely the cognitive import of any vital metaphor to any set of univocal, literal, or non-tensional statements. For special meaning, and in some cases even a new sort of reality, is achieved which cannot survive except at the intersection of two perspectives which produced it (pp. 243-244).

These two perspectives are similar to what Koestler (1964) identified as the interaction of the conscious and the subconscious mind in the act of creation pertinent to the scientist and the artist. Interestingly enough, the scientist, according to Koestler, seems drawn to emphasize the role of visualization, as opposed to the Aristotelian syllogistic reasoning, to account for the creation of science. The artist, on the other hand, quantifies creativity as a process, not through the mathematical, usually, but through verbal expression, so in this sense the creative process is stereoscopic. To illustrate this idea, Koestler uses the well-known urn/faces print whose doppelgänger nature is revealed through stereoscopic shifting (see Figure 1).

On the other hand, Berggren (1962/1963) also notes that "the most serious and interesting danger is that a given metaphor or allegorical expression may be transformed into a myth" (p. 244). He does not elaborate, but he commits the fallacy of dividing and classifying metaphors as different types of "tensive symbols." Fortunately, he arrives at the idea that metaphor cannot be translated into the literal, and he disparages the case studies Beardsley advocates. "Positivism," he maintains, "must ultimately admit defeat in both areas" (p. 250). Unlike Koestler, Berggren's theory does not approach an epistemology of metaphor. Furthermore, it might be questioned as to whether to become myth means to become allegory or if becoming myth means, as Nietzsche (1989, 1990) would argue, that the idea as metaphoric has been merely forgotten. As a myth, the persistence of the SSA as an example is noteworthy. It still exists, as I explore in greater detail in chapter four, in secondary-school textbooks, where it serves as a pedagogic device. So is it a myth? No better metaphor has replaced it.

Figure 1. Faces or urn? (EyeSearch.com, 1998).

For the purpose of chapter four, it is also worth noting what Berggren (1962/1963) has to say about Kelvin and Maxwell. According to Berggren, when Maxwell could not (or would not) create a traditional physical model for his electromagnetic theory of light, Kelvin criticized him over this point, which means that Kelvin, according to Berggren, nearly credited models with the same weight as theories, when a cursory examination of Kelvin's work reveals that he did in fact equate model and metaphor with theory. Models themselves behave analogically and metaphorically as they pass from one realm of thought to another. However, Berggren sees the danger as when

> univocal identification of poetic schemata with non-spatial reality produces poetic myth, so any naive fusion of scientific models with scientific theories is also one source of scientific myth. This form of science myth occurs . . . when what begins as an imaginative construct, used to construe a theory, gradually becomes identified with the theory itself and even assumes an independent and substantial reality of its own (1962/1963, p. 455).

Berggren does not deny science its metaphors, however. Instead he recommends "stereoscopic vision" as the solution to the dilemma (1962/1963, p. 456). Stereoscopic vision allows the scientist to maintain an equilibrium between, and a consciousness of, metaphor and reality, but such a notion again suggests there is a use of language that is nonmetaphoric.

The tensionists, then, brought to the study of metaphor an approach that began to examine how metaphors work. Their curiosity is analogous to why humanity has sought to explore outer space in the sense that, military and scientific objectives aside, we simply want to expand epistemologically. Such an intention fueled NASA for many years before it was realized that space travel technology could become part of an effort to divert an asteroid on a collision course with Earth. For metaphor, what is out there, or in this case, in the metaphor? This question bears answering as the extent to which we interact with science and technology increases. Beardsley (1962) is Aristotelian in his sense of wonder when he examines the tension point between the subject and predicate; however, he is more interested in the metaphorical moment and how metaphors work. Berggren (1962/1963) presents the idea of stereoscopic vision, the idea of seeing and not seeing. A greater consciousness of how metaphors work can also inform how metaphor might be taught to scientists. More direction can be found in that regard in the work of the interactionists, who took the study of metaphor a step further.

THE INTERACTIONISTS

Max Black (1962) contributed to the study of metaphor by promoting the idea of interaction. Being able to interpret metaphor depends upon not only the word but also the context of the sentence as a network. As an example, Black uses the often cited, "Man is a wolf," which invokes a simple metaphor expressed through a linking verb. Black calls the "focus" of this sentence the point where the metaphor occurs. The "frame" is the rest of the sentence. Focusing on these distinct sentence elements allows concentration on the metaphoric word without trying to fix it to a specific definition, and it goes beyond Richards as well as Perelman and Olbrechts-Tyteca and their naming of a metaphor's parts, which is really just another form of division and classification. Such an integrated way of thinking about metaphor is called the "interaction theory." Black's simple example strips metaphor to its essentials, something that the scientific metaphor does anyway, another reason why it is worthy of discussion.

The verb, even a simple one such as "is," provides the interaction, so in a sense, the verb itself is metaphorical. The interaction becomes both an "is/is not." The verb creates the tension/motion that underlies the Aristotelian notion of *energia*.

According to Ricoeur, Black contributed to the theoretical study of metaphor in three ways. First, metaphor depends not only upon the word but also upon the context of the sentence. For example, with "The chairman plowed through the discussion," some words are used metaphorically while others are not. The idea of the "focus" and the "frame" apparent in this example is the interaction theory in a nutshell.

Second, Black classifies the interpretation of classical theory as falling in one of two camps: comparison or substitution. Richards (1936) and Perelman and Olbrechts-Tyteca (1969) can be thought of as representing comparison theory because of their stab at a theory that breaks down the parts of metaphor, while Weaver (1990) and the author of *Rhetorica ad Herennium* represent substitutionists in the classical sense because the *Rhetorica ad Herennium* offers guidance on how to use metaphor, while Weaver's approach, though epideictic, is somewhat intuitive as it makes its leap of faith.

Third, Black (1962) questions why there is no notion of why some metaphors work and others fail. The same metaphor posed in different languages shows that metaphor is not bound by syntax or semantics.

Richard Boyd (1993) understands Black's metaphors as open-ended. On the subject of dead metaphor in the sciences, he argues that science serves as a semantic source, especially where theory is concerned. Metaphor, according to Boyd, creates an opportunity for science to accommodate language to the world, "arranging language so that it cuts the world at the joints" (p. 483). He notes the role metaphors may play in pedagogy and specifically cites the SSA. For Boyd, what is important is the social aspect of metaphors, for the way they become social entities as they are passed from scientist to scientist, an aspect that is not apparent with literary metaphors. Hence, they become "the property of the entire scientific community, and variations on them are explored by hundreds of scientific authors without their interactive quality being lost" (p. 487). Such a distinction makes them more interesting to study than literary metaphors in terms of learning how they are used, since scientific metaphors are usually stripped to the bare essentials, and they are used epistemologically, as opposed to only experienced, by a variety of audiences.

It is appropriate to follow a discussion of Boyd (1993) with a discussion of Kuhn (1993), since he has commented upon Boyd's concept of "cutting the world at the joints." On Kuhn, Boyd has observed, "[His] work has made it clear that the establishment of a fundamentally new theoretical perspective is a matter of persuasion, recruitment, and indoctrination" (p. 486). While Kuhn agrees with Boyd's basic interpretation of Black, he differs in terms of his approach to the idea of "cutting the world at the joints." Specifically, Kuhn disagrees that language can ever "cut the world at the joints." Boyd has cited the historical development of language and science as evidence of his claims. Kuhn concedes Boyd's point that earlier languages might have more accurately portrayed the world in terms of a thing-to-object correspondence. However, such an admission more readily supports a substitution view of language than Black's (1962) interactive quality expressed by metaphor. According to Kuhn, Boyd's point is that "nature has one and only one set of joints to which the evolving terminology of science comes closer and closer with time" (p. 541). In that sense, Boyd's approach is more similar to that of nineteenth-century Scottish Natural Philosophers and the ancient Greeks, who would discover how nature's puzzle fits together, but Boyd's work can also be read as substitution theory, according to Kuhn's criticism. Kuhn concludes that to the contrary, science today, with its greater dependence upon instruments that record what cannot be directly observed, is "more . . . Aristotelian than . . . Newtonian" (p. 541).

Of course, there are other reasons to consider Kuhn in relation to this discussion. His *Structure of Scientific Revolutions* (1970) is essential to any discussion of how the SSA developed in the work of these physicists as well as the contemporary biologists. In these cases, metaphor is a social construction.

The interactionists, then, placed more emphasis on metaphor's philosophical significance. With them, metaphor becomes a social construction as the discussion moves ever closer to the sciences. In this discussion, strands of the substitutionists are apparent in the work of Boyd (1993), yet he also adds to the dialogue by bringing metaphor into the realm of social construction. The final step remains to be taken into more recent developments into metaphor as an epistemology. Though metaphor's value as an epistemological device can be traced back to Aristotle, as I have demonstrated, its value has attracted fresh interest from artificial intelligence researchers who would technologically recreate the human mind.

METAPHOR AS EPISTEMOLOGY

With the Solar System Analogy, the epistemological web stretching from knowledge of the solar system must be considered; indeed, this web becomes richer as the implications of subatomic particles as planets are borne into the discussion. Such a metaphorical leap draws speculation upon orbits and the gravity required to hold the subatomic particles in sway. Then, the type of relationship these particles have with a central point, the nucleus, must be discerned. Is there an analogous relationship between this system of subatomic particles and astronomical comets that visit a solar system? What is the relationship of this atomic system to other systems and to the greater whole, just as in astronomy the relationship of the solar system to the galaxy and to the universe might be questioned? Clearly, interaction theory has led to a rich network of ideas for the scientific metaphor.

The SSA renders a great significance in the sense of allowing access to Black's (1962) "network of meaning." However, there are other theoretical approaches. One is Karl Popper's (1972) concept of falsification that does not name metaphor but alludes to it and to rhetoric in general, ultimately leading to epistemology.

Popper (1972) begins his discussion of falsification by positing the difference between science and psuedo-science, by which he means astrology, but he is also thinking of psychology and Marxism. The problem is that once people become convinced that one of these "sciences" is correct, they see testimony to its verity everywhere. According to Popper, these sciences contrast sharply to Einstein's theory of gravitation in the sense that through his theory, Einstein specifically predicted how gravity affects celestial light. These predictions were then verified empirically when observation revealed light to be affected by heavy astronomical bodies.

Psychology and Marxism, according to Popper (1972), share more with myth than with science. To some extent, this common ground can be seen as metaphoric. In his discussion, there is an undercurrent of the rhetorical as well, with specific applications to metaphor.

For Popper (1972), the psuedo-sciences operate out of a more ancient tradition that suggests myth as rhetorically substantiated. In general, he is interested in discerning points of "demarcation . . . a criterion of the scientific character of theories" (1972, p. 136). However, Popper links myth and science because "myths may be developed, and become testable; that historically speaking all—or very nearly all—scientific theories originate from myths, and that a myth may contain important anticipations of scientific theories." He then cites examples of "Empedocles' theory of evolution by trial and error, or Paremenides' myth of the unchanging block universe in which nothing ever happens and which if we add another dimension, becomes Einstein's block universe" (1972, p. 134). It is important to note here that Popper does not completely discount myth, just as he does not discount Marxism or psychology.

As a grounding in philosophy, Popper discusses David Hume's idea that there is no reason to deduce that because one instance occurred that a similar one will follow. Such an assumption would have to be traced back *ad infinitum*. Hume (1964) offers remediation of these views by pointing out that laws are established by continuous association of events. Popper counters that such a solution is psychological rather than philosophical because it provides only a psychological basis for belief in laws. Popper (1972) amends Hume by noting that his approach allows for not merely describing life but hypothesizing it through "repeated observation" (p. 138).

Popper asserts that Hume errs on three points: "(a) the typical result of repetition; (b) the genesis of habits; and especially (c) the character of those experiences or modes of behaviour which may be described as 'believing in a law' or 'expecting a law-like succession of events'" (1972, p. 139). Popper then expands upon these objections:

- Repetition becomes abbreviated. As an example, Popper cites playing the piano. What begins carefully becomes an automated physical response destroyed as a conscious act by becoming as superfluous as the metaphoric aspect of a word as it passes into language and common usage.
- Habits, such as eating, do not necessarily begin with repetition, but rather out of uncontrollable need. The same argument might be made for language in general. Parents are relieved, for example, when a child becomes old enough to specify the source of pain and discomfort.
- Believing a law is not the same as behavior, because with a law, an expected chain of events must transpire. Metaphor, if we accept it as at the seat of language, is behavior rather than rhetoric.

To interpret Hume coherently, it must be assumed that the events whose observation create laws are not identical but similar, which

> means that, for logical reasons, there must always be a point of view—such as a system of expectations, anticipations, assumptions, or interests—before there can be any repetition; which point of view, consequently, cannot be merely the result of repetition (Popper, 1972, p. 140).

A point of view is necessary for metaphor as well. Such a view on perspective can be thought of as stereoscopic, for example, as Berggren (1962/1963) has posed.

Popper (1972) notes that "Without waiting, passively, for repetitions to impress or impose regularities upon us, we actively try to impose regularities upon the world." These regularities can be read as how metaphors are created to explain the unknown with the known. He continues with, "We try to discover similarities in it, and to interpret it in terms of laws invented by us." In a sense, the metaphors constructed for scientific explanation are laws as well, such as the SSA. He concludes with, "Without waiting for premises, we jump to conclusions. These may have to be discarded later should observations show that they are wrong" (1972, p. 142). The discarding of observations is similar to what occurs with scientific metaphors. However, they are not always discarded, but shift shape as they morph into new incarnations, as the SSA continues to be used for education.

For Popper, the process of "trial and error" becomes the "conjectures and refutations" condensed to "falsification."

Observation is an important part of the process of falsification. Observation, Popper notes, "is always selective" (1972, p. 143). That selection process is metaphoric in itself and requires what Popper refers to as "a descriptive language, with property words; it presupposes similarity and classification, which in its turn presupposes interests, points of view, and problems." These property words are the very stuff of metaphor. As examples, Popper notes that when an animal needs to feed, its world becomes what it can and cannot eat. When it is in danger, its world becomes escape routes and hideouts. He concludes that, "We may add that objects can be classified, and can become similar or dissimilar, only in this way—by being related to needs or interests" (Katz in Popper, p. 143). Such are the choices that must be made when a metaphor is used, or disregarded, or in the process of interpreting a metaphor, when what is false is ignored. Just as Hume's (1964) justification of induction is psychological, so is Popper's assessment of the human search for the similar and dissimilar.

The danger in the search for the similar is in what Popper refers to as dogmatic thinking:

> We expect regularities everywhere and attempt to find them even where there are none; events which do not yield to these attempts we are inclined to treat as a kind of 'background noise'; and we stick to our expectations even when they are inadequate and we ought to accept defeat (1972, p. 145).

The problem, then, for metaphor, is the tendency to adhere to it when doing so is no longer useful and can even be counterproductive. This point is where the value of falsification must be remembered, because it is at the seat of metaphor. An atom is not a solar system, but it must not be *forgotten* that an atom is not a solar system. On the dogmatic attitude, Popper comments further that it

> is clearly related to the central tendency to verify our laws and schemata by seeking to apply them and to confirm them, even to the point of neglecting refutations, whereas the critical attitude is one of readiness to change them—to test them; to refute them; to falsify them, if possible. This suggests that we may identify the critical attitude with the scientific attitude . . . (p. 147).

However, Popper does not pose falsification as a replacement for the dogmatic, but rather as "superimposed upon it: criticism must be directed against existing and influential beliefs in need of critical revision—in other words, dogmatic beliefs" (p. 147), so falsification enhances rhetoric.

Popper alludes to rhetoric in the sense that the defense of myths is rhetorical, and science can become myth, especially when science is represented by metaphor. On the ancient Greeks, he notes their critical method gave rise to the mistaken hope that it would lead to the solution of all the great old problems; that it would establish certainty; that it would help to prove our theories, to justify them. But this hope was a residue of the dogmatic way of thinking; in fact, nothing can be justified or proved (outside of mathematics and logic) (p. 148).

Here Popper alludes to the rhetorical as a fault in the ancient Greek reasoning when it was applied to science. However, he fails to recognize that mathematics and logic consist of metaphors and are therefore rhetorical.

Popper's discussion, then, touches upon the rhetorical, with suggestions for the metaphorical. More specifically, he focuses on similarities and comparison to how the world is shaped through the process of falsification. The defense of myth is rhetorical, and critical examination does not counter the rhetorical stance so much as it complements it.

Arbib and Hesse (1986) explore myth as a way of creating knowledge. They agree with Black's (1962) interaction theory, especially when L. Wittgenstein's (1958) family of resemblances is considered, which suggests that the language of observation will be rife with predetermined classifications that suggest images that cannot be expressed verbally. Gilbert Ryle (1951) has concurred with Wittgenstein's refusal to tie thought to language or to image. The image itself is deceptive since so much in the metaphor must be ignored for it to function effectively: "The recipient of the metaphor [drawn between a thinker and a fruit picker] is expected to discount the concrete details of ladder, branches, right arms, and apples, without which there would be no apple-picking" (Ryle, 1951, p. 67), and the same problems might be said to apply to imagistic thinking as well. The mountain guide, for example, may peer up at the mountain through a telescope and plan a climb from the hotel, but he may not be able to articulate a route. Therefore, language cannot be analyzed in terms of comparing the language experience with that of the world because language is biased by theory. On the other hand, imagery suggests fixed meaning. Hesse (1970) has observed that science has been altered in the past to fit theory, and she wonders to what extent such a practice continues today. The SSA is a good example of the attempts to make subatomic physics fit within the scope of classical mechanics. Its ultimate dispensation indicates what others have noted about metaphor: that it is disposed of when no longer needed; yet as I have noted and will explore in more detail, it persists in many ways.

To what extent is the metaphor really disposed of? Certainly it passes into literal language, and the description alters somewhat as new metaphors are picked to describe the phenomenon. In the case of the SSA, it has passed into science textbooks. Some maintain the SSA, but others use the plum pudding metaphor, the beehive metaphor, or other astronomical models, such as the analogy between Saturn and its rings (a much less accurate one since it suggests subatomic particles in planar orbits).

Arbib and Hesse (1986) observe that science and religion rely to a certain extent upon metaphor because both must describe that which cannot be directly observed. At this point, they arrive at schema theory, which is similar to Black's (1962) pattern of interaction, with the addition of a theory that can build on theory to make predictions. Arbib and Hesse define a schema as, "Instead of thinking of ideas as impressions of sense data, we visualize an active and selective process of schema formation that in some sense constructs reality as much as it embodies it" (p. 43). The question addresses how knowledge is constructed. They pose the interaction as a schema, which is

> both a process and a representation. The formation and updating of the internal representation, a schema assemblage, are viewed as a distributed process, involving the concurrent activity of all those schema institutions that receive appropriately patterned input (p. 54).

Metaphor is relevant because, "language [is] a mediation between schema assemblages that differ" (p. 62). They relate these ideas to scientific theory as an epistemology because individual schemas alter, so "most of us would agree that there is an external spatiotemporal reality independent of human constructions and providing the touchstone for our attempts to build physical theories" (p. 62). The physical theories that interest Arbib and Hesse relate to artificial intelligence.

Recent research into artificial intelligence has focused on how computers can layer symbolic language so that its usage becomes metaphoric. To arrive at such a goal, the metaphor's interactivity must be recognized, which recalls Black's (1962) approach to the concept. In addition, artificial intelligence must recognize the movement from McMullan's (1976) U-fertility (unknown-fertility) to P-fertility (proven-fertility), which means that artificial intelligence must be able to use metaphor as people do, to propose metaphors and then follow up on their verity. Black's (1962) interaction can be read here because the interaction of knowledge and language results in the creation of theory through metaphor. According to Arbib and Hesse, "both schema theory and the network view of science have led to a theory of language in which metaphor is normative, with literal meaning as the limiting case" (1986, p. 171). Such an observation flies in the faces of substitution theory, which treats metaphor's defining moment as its deviation from accepted meaning. Even Black (1962) believes some words to be used metaphorically, but others not. According to him, if all of the words in a sentence are metaphorical, then the use is a "proverb, allegory or riddle (p. 27) and that "metaphor plugs the gap in literal vocabulary" (pp. 32-33). Arbib and Hesse would disagree.

On a similar note, W. H. Leatherdale (1974) has weighed in on myth, which he considers the paradox of literalness versus the metaphorical. Leatherdale cites Van Steenburgh because he "holds that metaphorical terms acquire meaning in a given context only by transference from the literal (i.e., ostensive) meaning" (p. 188). He recognizes that the term "ostensive" is vague, as does Van Steenburgh, and recommends that criteria be established for its recognition.

Leatherdale (1974) further recognizes the limits of ostensivity when he notes that it "is not co-extensive with sense-data . . . I incline to think it is co-extensive with what is perceptible or what is 'directly' perceptible" (p. 188). To illustrate, Leatherdale poses the idea of the witnessing of an honest act as an example of honesty. However, the further removed from the witnessing of such an act, the greater is the reluctance to assign honesty as an accurate descriptor, such as with honesty as an example of character or with an aphorism such as, "Honesty is the best policy." An extremely important aspect of this problem is illustrated by "positivists in their abortive attempts to reduce all science to phenomenal elements of some kind" (p. 189). Leatherdale's answer is to establish criteria for ostensivity (p. 190).

At first glance, such an approach would seem to pay too much homage to classical substitution theory and to be a misapplication of Aristotelian theory. However, the question Leatherdale addresses is how to recognize that metaphor has become myth. Failing to recognize this can negatively influence science, and it is indeed a tightrope that users of metaphor must walk.

In her earlier works, Hesse (1970) proposed a historical approach, which is a reasonable idea, for, as McMullan noted, the idea of the "historicist turn" is "to say that the unit for appraisal on the part of the working scientist is not a theory considered as a timeless set of propositions . . . [because] the theory taken over its entire career to date . . . impose[s] upon the scientist the task of historian" (1976, p. 691). Such a task is one Hesse explores through setting up a debate between a Campbellian and Duhemist. This debate is indicative of how a historical approach can shed led light on contemporary problems.

Pierre Duhem was an eighteenth-century physicist who derided the British for their use of analogy to develop theory. N. R. Campbell was a British physicist who addressed the Duhemist argument in his 1920 *Physics: The Elements*. The Duhemist argument, according to Hesse, is that, "the use of models or analogues is not essential to scientific theorizing," much less metaphors and analogies. Theory can be explicated through deductive reasoning, and the results can be tested through experiment and observation. However, a chink in the Duhemist armor occurs when model is briefly admitted, only to be dispensed with the arrival of theory (Hesse, 1970, p. 7).

The Campbellian counters by first posing that not only is model valuable, but specifically, analogy, which can be divided into three distinct types: positive, neutral, and negative (Hesse, 1970, p. 8). The Campbellian defines a model as "any system, whether buildable, picturable, imaginable or none of these, which has the characteristic of making a theory predictive" (Hesse, 1970, p. 19). The Duhemist objects to the fact that a mathematical model is deduced from observation, and it is at this point that the Duhemist encounters problems with the discussion of theoretical versus observational terms. The Campbellian advises that, "you must allow for the frontier between them to shift as science progresses" (Hesse, 1970, p. 23) and concedes that a model might fail. For example, to explain motion, an analogy can be drawn between balls on a pool table. For such a discussion, the color of the balls would not be important, but their velocity would be. The concept of falsifiability is endemic to scientific theory since it

> is required to be falsifiable in the sense that it leads to new observation statements which can be tested . . . [and] that it leads to new and perhaps unexpected and interesting predictions. But here there is an ambiguity. The weaker sense of such a requirement is that new correlations can be found between the same observations' predicates; the stronger sense is that new correlations can be found which involve new observation predicates (Hesse, 1970, p. 37).

Falsifiability is more relative to metaphors since they are easily shifted and easily disproved. When disproval occurs, they are dispensed with, but the fact that they serve for dispensation indicates their value. They allow for a science that admits intuition or probability. Quantum physics may be cited as an example of when metaphor has been discarded for the sake of quantitative expression, but it must be remembered that mathematics is simply another language, and Richards' (1936) proclamation that "All language is metaphoric," (a statement with which Arbib and Hesse agree) should not be read as confined to natural language. Perhaps the idea that "All language is metaphoric" should be amended to "All languages, natural and artificial, are metaphoric."

Analogies with classical physics are drawn as a way of comparing, contrasting, and therefore explaining, quantum physics. The SSA works in this fashion. As another example, in terms of the metaphor drawn to explain the structure of light, the particle metaphor's strengths dovetail with the wave metaphor's weakness, and vice versa. If only the positive aspects were drawn, the boundaries of knowledge would be ignored. The false aspects reveal those boundaries and expand such study epistemologically (Hesse, 1970).

Hesse by no means accepts metaphor and analogy without some reservations. To the contrary, she clearly regards them as useful tools for which she lays out specifications:

> If a model is to be scientifically useful, it must be in itself familiar to us, with its laws well worked out, and it must be easy to extend and generalize it so that its other properties, which we have not so far used, may be related, if possible, with the other properties . . . The model must have as it were, a life of its own (p. 171).

She also discusses how models might be used. A model might serve to propel a scientist to the next level of thought, and then it might be omitted from further consideration because it no longer serves a purpose. The SSA certainly has served this purpose.

Other analogies are disposed of and then resurrected because they are valuable. The wave and particle descriptions of light are good examples. Descartes thought of light as a wave. On the basis of his observation of the prism, Newton discarded Descartes' wave theory and described light as a particle. However, Young's double slit experiment led him to believe that light is a wave. Today, light is thought of as having wave properties and particle properties. These analogies serve to illustrate what Hesse means when she writes, "Analogies like these . . . have enormous vested interest in a theory, and therefore they can never easily be abandoned when new facts do not appear to fit in with the system of explanation which the analogies presuppose" (1970, p. 174). Of course, such an attitude leads to problems in which the Duhemist would revel since, according to Hesse, "There is always the temptation to explain awkward facts away in order to save the basic analogies of a science" (1970, p. 174). Hesse refers here to how scientists ignore contradictions and other problems that do not fit the current paradigm because the scientists are either blinded by the metaphor or have too much invested in it to look for a new one. The prior discussion of the nature of light illustrates how a metaphor is dispensed with and then resurrected to direct theory. It is important to be aware, however, that metaphor can be dispensed with in such fashion, and in fact should be, if the data warrants it.

Ina Loewenberg (1973, 1975) has written of the epistemology of metaphor. She is especially concerned with the idea of truth as it applies to metaphor. If you say, "Sam is a plumber," when he's really a doctor, what are you saying about Sam? To what extent is it true that Sam the doctor is a plumber? Loewenberg focuses on accounting for "how metaphors can be understood, identified, and assessed" (1973, p. 31). Her concern is with what she refers to as a "novel" metaphor, not a dead one whose truth has been realized. Her "preference is to take metaphors as true merely by stipulative definition when novel and as no longer metaphorical, but literally true or false when deceased" (1973, p. 41). The importance of the truth-value, according to Loewenberg, is that it positions metaphors where the positivist cannot ignore them "and rescues them from the purgatory of emotive meanings or the limbo of meaninglessness" (1973, p. 41). Like Ricoeur (1975), Loewenberg (1973) values the semantic extension afforded by metaphor, and within the broader expression of analogy, something occurs that passes beyond the concepts of "filtering," "interacting," or "stereoscopic vision." Such an observation harkens to Aristotle's "nameless act."

Loewenberg (1975) has also written of the dead metaphor issue. To explicate her thoughts, she discusses the paradox of comparison with the idea of open versus closed comparisons as an avenue of discussion. If a metaphor is open, according to Loewenberg, then it is vacuous, because it can be interpreted in too many ways to be meaningful. If it is closed, then it can be paraphrased, or interpreted, which Loewenberg finds too mechanical and requests of the metaphor too much complexity of design and strength. What is lost through interpretation is not easily resurrected, and such an approach does not fit with different types of metaphoric usage. For example, simple metaphors such as "Light is a wave" can, at first glance, be easily interpreted, and so can "An atom is a miniature solar system," but either can be expanded to analogy, so at what point does the analogy stop? To what extent does gravity affect light? Comets visit solar systems. What, in the SSA, is analogous to comets? Furthermore, analogies are easily interpreted by an audience (though perhaps not in the way they were intended). For these reasons, Loewenberg asserts, metaphors are not comparisons.

To determine what questions are answered by physical models, McMullan (1968) suggests to first ask, "What do complex postulated structures of the scientist tell us of the world?" McMullan notes that the scientist begins with ideas isolated for exploration within a specific realm of inquiry. Theory is never intended to reify what is empirically known, but to reach for some type of structure to articulate. McMullan perceives the model as an entity that exists apart from natural or mathematical incarnations. If an atom is a miniature solar system, then as a model, it exists as a natural language expression, and it can be expressed mathematically in terms of orbits and trajectories and so forth. However, whether in the natural language or mathematical incarnation, it may become too open or too closed, as Loewenberg (1975) has noted. For McMullan, the model is an entity unto itself, and it does not so much reflect the linguistic or mathematical interpretations as it exists separately from both. McMullan notes that "Bohr's theory is about hydrogen atoms, and the statements comprising it make use of terms like 'electrical charge,' [and] 'electron,' which prevent it from also describing planetary systems." Therefore, "the physical theory makes an assertion about a physical sub-structure which can account for data; the phenomenological model makes no assertion" (p. 391), and it is the phenomenological model that accounts for theory, not the reverse. It also worth noting that McMullan refers to "the Bohr model of the atom" as "one of the two most productive models of our century" (1968, p. 392) though he does not mention what he believes the other one to be. Perhaps he is thinking of the analogy of the travelers at the train station that illustrates Einstein's theory of relativity.

McMullan's idea of the primary importance of metaphors, that of their fertility, has been briefly mentioned, and this concept bears further development. He first expounds upon the idea by noting that the creation of a new theory is analogical, that the scientist examines what is known and asks, "What if?" For the theory of atomic structure, W. Thomson (1910) and J. C. Maxwell (1986a, 1986b) began with the ancient Greeks' idea that the atom was indivisible. Though their attempts to explain how the atom contributes to molecular structure are regarded today as without value, their vortex theory allowed the SSA to develop. What the scientist seeks is the concept of proven fertility (P-fertility) "that confirms the truth-value of a theory not its as-yet-untested promise (U-fertility)"

(McMullan, 1976, p. 686). The value of the metaphor as a model, according to McMullan, is the extent to which the scientist travels along it. Included in the accounting of the trip are the dead ends as well as the deviations necessary for the model to continue to be useful. Though Thomson and Maxwell's work was in one sense forgotten upon the discovery of subatomic particles, the SSA continued to contribute epistemologically. With chronological perspective, the scientist can judge which deviations were indeed fruitful as a way of determining the *P*-fertility. Lest it seem that the model that most closely constructs theory be considered stronger in terms of P-fertility, McMullan concludes that "The novelty and the variety of the predictions a theory generates are significant first because they subject the theory to severer tests, an important desideratum if corroboration be proportionate to severity of test" (1976, p. 693). Therefore, the fact that a theory can be falsified should not in itself be considered the only way to judge it. To the contrary, according to McMullan, the more ways in which a theory can be falsified, the more fertile it should be considered.

H. G. Petrie and R. S. Oshlag note that metaphor typically is misjudged as an educational tool: "Metaphors are used when one is too lazy to do the hard analytic work of determining precisely what one wants to say." As proof, they cite a study that concluded that "very common and useful analogies and metaphors used in the instruction of physicians come to interfere with later learning and a more adequate understanding of the concepts" (1993, p. 581). However, they recognize that "theory-constitutive metaphors . . . are integral parts of the very structure of a theory at any given time in its development" (1993, p. 581). They refer to dead metaphors as "residual metaphors" that can serve to instruct. Paradoxically, what is metaphoric to the physicist may be literal to the physics student. For example, the idea of light as a wave or as a particle has different meanings to the physicist than to the student in an introductory class:

> If, however, we insist . . . that learning must always start with what the student presently knows, then we are faced with the problem of how the student can come to know anything radically new. It is our thesis that metaphor is one of the central ways of leaping the epistemological chasm between old knowledge and radically new knowledge (1993, p. 583).

The analogy between the solar system and the atom is comparative (and quite literally dead) to the teacher but interactive for the student. To conclude, they refer to Kuhn and the way a child learns to differentiate between ducks, swans, and geese. The child has no internal rules of classification but rather learns to differentiate by comparing and contrasting (1993, p. 588).

George Lakoff and Mark Johnson's *Metaphors We Live By* (1980) is an often-cited work in metaphor studies. As a compromise to the substitution or correspondence theory of metaphor, they propose an experientialist perspective that is a correspondence theory in the sense of focusing on a truth value for statements that correspond with what has been witnessed in the world. The idea of a truth value depends upon a literal meaning for words, however. They deviate from correspondence theory in the sense that they recognize that there may be more than one truth, and that truth is a relative term. The problem with metaphor, according to Lakoff and Johnson, especially in the sciences, is that Aristotle's admiration of metaphor did not survive the seventeenth-century scientific revolution: "The fear of metaphor and rhetoric in the empiricist tradition is a fear of subjectivism—a fear of emotion and imagination. Words are viewed as having 'proper senses' in terms of which truths can be expressed" (p. 191). The experientialist position "unites reason and imagination" (p. 193). Reason is united in the sense that a metaphor calls upon the ability to categorize, and the imagination provides the ability to perceive one thing as another. The fusion, for Lakoff and Johnson (1980), is "imaginative rationality" (p. 193).

On the subject of dead metaphor, they argue that some instances such as "*foot* of the mountain are idiosyncratic, unsystematic, and isolated. They do not interact with other metaphors, play no particularly interesting role in our conceptual system, and hence are not metaphors we live by" (p. 55). Such a dead metaphor is contrasted with others such as "wasting time, attacking positions, going our separate ways," which they argue are "reflections of systematic metaphorical concepts that structure our actions and thoughts. They are 'alive' in the most fundamental sense: they are metaphors we live by. The fact that they are conventionally fixed within the lexicon of English makes them no less alive" (p. 55). Though I would argue that these metaphors are equally dead, for Lakoff and Johnson (1980) they retain *energia* because of how they are categorization metaphors that interact with many others.

CONCLUSION

These views of metaphor as they are applied to science and to rhetoric in general have enriched the dialogue. First, Aristotle is credited with philosophizing metaphor. He also introduced what became the substitution or comparison theory. However, this approach was too limiting. Nietzsche then named all language as metaphoric, which, if it did not directly influence many twentieth-century rhetoricians and philosophers, was symptomatic of the discomfort with the modern and postmodern interpretations of the Aristotelian metaphor. Beardsley finds the tensionist theory interesting, but his idea of the metaphorical twist, though somewhat syntactic, arrives at a tensionist semantics. Black furthers Aristotle's idea of "bringing before the eyes" as it applies to metaphor by explicating the interaction theory. Arbib and Hesse would extend Black to an epistemology of metaphor. Clearly, these theorists contribute to understanding the specific instance of the SSA in its network of ramifications, which belie the way in which it has been perceived as an interesting but no longer useful artifact. These considerations further inform examination of metaphor in the context of contemporary and nineteenth-century scientific writing as well as considerations for technical communication texts and the technical communication classroom.

Two questions arise from this discussion that bear further examination. First, there is the idea that there is a use of language that is not metaphorical. Most of these rhetoricians and philosophers, except for Nietzsche and Hesse, agree that there are words that are not metaphoric. For Richards, language is least metaphoric in the "settled sciences"; for Perelman and Olbrechts-Tyteca (1969), the metaphor must be dismantled like scaffolding as scientific investigation becomes law. Others, such as D. Berggren (1962/1963), favor a "stereoscopic vision" that allows us to see simultaneously the literal and metaphoric; Boyd (1993) proposes that Anglo-Saxon words might once have "cut the world at its joints." How can all language be metaphoric if there is a literal use of language upon which the metaphor is based?

The other question concerns mathematics. Richards (1936), Black (1962), and Perelman and Olbrechts-Tyteca (1969) have proposed that mathematics lacks metaphors. Is this true? If it can be successfully argued that mathematics is not metaphorical, then there would be greater weight for an argument that there is a literal use of natural language that is not metaphorical. If mathematics is metaphorical, then the case for a natural language that is not metaphorical would be diminished.

These ideas are examined in more detail in the next two chapters. In the next chapter, I consider the idea of whether or not mathematics is metaphorical. In the chapter following that one, I examine the question of natural language as metaphorical.

CHAPTER 4

The Metaphors of Mathematics:
A Case Study of the
Solar System Analogy

Before we can settle whether or not all language is metaphoric, it would be a good idea to examine whether or not mathematics is metaphorical. A number of rhetoricians and philosophers, such as Richards (1936), Perelman and Olbrechts-Tyteca (1969), and Black (1962) have proposed that mathematics lacks metaphors. If mathematics is a language, albeit an artificial one, is it free from metaphors? Have human beings, whose language is metaphorical, succeeded in creating a language that is not metaphorical? Perelman and Olbrechts-Tyteca further suggest that mathematics is not metaphorical when they write, "If the analogy is a fruitful one, theme and phoros are transformed into examples or illustrations of a more general law, and by their relation to this law there is a unification of the theme and the phoros" (p. 396). Does mathematics represent the unity of theme and phoros since physical laws are often described mathematically? Is Black correct when he says that what begins in metaphor ends in mathematics? Does this mean, then, that mathematics is not metaphorical? On analogy, Perelman and Olbrechts-Tyteca (1969) have further claimed

> that what distinguishes analogy fundamentally from simple mathematical proportion is that in analogy the nature of the terms is never a matter of indifference. For the effect of analogy is to bring the terms A and C and B and D closer together, which leads to an interaction and, more specifically, to increasing or decreasing the value of the terms of the theme (p. 378).

Perelman and Olbrechts-Tyteca posit here that there is a different type of interaction in an analogy than in mathematics, as if the combining of the terms generates an abstract field of metaphorical energy that they see as the value of metaphor. Are mathematical terms those of indifference, though? Are the terms indifferent to one another? What does it mean for a term to be "indifferent?" Does the number "5" differ in terms of meaning from the word "five?" Is "five" an example of a word that is not metaphorical? As to increasing or decreasing the value of theme, it would first need to be determined if mathematics is metaphorical.

On the other hand, Berggren (1962) has compared a physicist, when using mathematics to construct theory, to a mapmaker who creates a map for an imaginary land. When this mapmaker refers to an actual place, however,

> a second referent is thereby introduced, and the iconic sign becomes metaphorical. For whenever any mathematical equation is used to interpret the natural world, one of the 'referents' is immanent within the system of mathematical relations, while the other referent is considered transcendent to such a system. Consequently, any sign focus which has at least two referents fulfills this condition of metaphorical usage (pp. 238-239).

According to Berggren, then, the physical phenomenon that the physicist describes is metaphorical, but the mathematics is not. If mathematics is universal, though, then can it really exist without metaphor?

Since mathematics is an artificial language, it is reasonable to examine it for evidence of metaphor. If it is not metaphoric, then we can proceed to the question of whether or not all *natural language* is metaphoric. If mathematics is not metaphoric, then the argument for all language as metaphoric will be weakened. Such a case would indicate that even if natural language is metaphoric, an artificial language has been created that is not metaphoric, one created by human beings, which would suggest that a more perfect, considered language, one that has developed over a few millennia, is preferable to a language that has developed over a much longer period of time and with less governance than is found in mathematics. It will also suggest that a natural language could be refined so that it is no longer metaphorical. If mathematics can be shown to be metaphoric, then the argument for all language as metaphoric will be stronger.

Traditionally, mathematics has certainly not been thought of as metaphoric. However, coinciding with the advent of computers and the widespread use of metaphoric terms such as e-mail, World Wide Web (or is it the Internet, or is it the information superhighway?), there has recently been some discussion of whether or not mathematics is metaphorical. One of the most notable explorations is Lakoff and Núnez's *Where Mathematics Comes From* (2000). Lakoff, a linguist, is well-known in the field of metaphor studies for *Metaphors We Live By* (1980), co-authored with philosopher M. Johnson.

Lakoff and Núnez (2000) list criteria that they believe characterize the "Romance of Mathematics." A couple of the primary ones are worth examining for this discussion:

- Mathematics is an objective feature of the universe; mathematical objects are real; mathematical truth is universal, absolute, and certain.

Metaphors are universal in the human experience since all languages contain metaphors, and it is rare to encounter a culture without a creation myth, so metaphors are also in that sense absolute and certain.

- What human beings believe about mathematics . . . has no effect on what mathematics really is. Mathematics would be the same even if there were no human beings, or beings of any sort. Though mathematics is abstract and disembodied, it is real (p. 339).

The argument here is that mathematics does not need human beings to exist. Planets would still be spheres, and leaves and stems would be fractals. It could easily be argued that metaphor exists because human beings have brought it into existence. However, it can also be argued that the observation of physical phenomena and application of the mathematical to measure those phenomena have greater value when they can be extended through the comparative operation of the mind that is at its root metaphorical. History has witnessed this sort of mental operation many times, such as when Einstein applied the theory of gravity to light waves. Because it was comparative, that is, because he was comparing how light affects physical objects to how gravity affects light, his discovery at its base was metaphorical, and it was later proved by observation.

Lakoff and Núnez (2000) more specifically address mathematics as metaphor. They begin by establishing that mathematics is innate, citing research that indicates that four-day-old babies can differentiate between two and three items, and that counting is something primates and rats can do. Like Nietzsche (1990) and Hesse (1970), their claim for metaphor is that it is "not a mere embellishment; it is the basic means by which abstract thought is made possible" (p. 39). They back up such an assertion with a careful analysis of mathematics, beginning with basic arithmetic.

For example, Lakoff and Núnez (2000) break the foundations of mathematics as metaphor into two categories. The first consists of grounding metaphors such as addition, subtraction, and the concept of sets, which, again, infants can distinguish. From these grounding metaphors are built what they refer to as linking metaphors such as points on a line, which can also be understood as a metaphor springing from the measuring stick and from the "peculiar properties of our bodies" (p. 86). The addition and subtraction metaphors are the basis of algebra, and geometric figures are described with algebraic equations. Lakoff and Núñez build upon these concepts to arrive at a detailed case study of Euler's proof, of which renowned Harvard mathematician Benjamin Pierce once said, it "is surely true, it is absolutely paradoxical; we cannot understand it, and we don't know what it means. But we have proved it, and therefore we know it must be true" (as cited in Lakoff & Núñez, 2000, p. 383), a comment I find more reminiscent of poems I have read than equations.

As I have noted, metaphor and analogy have long been a part of scientific thought. A good example of a pedagogy that encouraged metaphor and analogy in university training may be found in Scottish Natural Philosophy, which flourished well into the nineteenth century as guiding pedagogy in the universities there. It is important to have an understanding of Scottish Natural Philosophy, because it influenced the scientists, specifically William Thomson (Lord Kelvin)[1] and James Clerk Maxwell, who developed the theory of atomic structure that led to the current conception of string theory, whose purpose is to formulate a unified theoretical view of the universe, a theory of everything. Their work influenced J. J. Thomson, Oliver Lodge, Ernest Rutherford, and Niels Bohr. The narrative history that follows in this chapter traces how the Solar System Analogy (SSA) was useful to these physicists. As the theory developed, the mathematics shifted with it, but the SSA was consistently a part of the development of this theory until Bohr dispensed with it in favor of a mathematical model, and quantum mechanics, which is yet another metaphor that describes what classical physics cannot. This study will follow the SSA as developed in the work of these scientists. First, I examine Scottish Natural Philosophy, which provided a foundational perspective for Kelvin and Maxwell.

SCOTTISH NATURAL PHILOSOPHY

Scottish Natural Philosophy is also referred to as the Common Sense school of philosophy, and it manifested itself during the mid-seventeenth century Scottish Enlightenment. Generally, Thomas Reid, James Oswald, James Beattie, and Dugald Stewart are regarded as its most prominent voices. According to these philosophers, everyone has access to the "common sense" that is the fount of human knowledge. For the Scottish Natural Philosophers, there was no doubt of the existence of concrete objects or of other people. Instead, Scottish Natural Philosophy valued human experience and denied that any type of analysis could undo the idea of true knowledge (Olson, 1975).

---

[1] William Thomson was the first scientist to achieve peerage as a result of his scientific work. His inventions related to the telegraph made him wealthy. When he was offered peerage, he picked "Kelvin" as his title from the Kelvin River, located close to the University of Glasgow (Sharlin, 1979). To avoid confusion with J. J. Thomson (to whom Kelvin was not related), William Thomson is henceforth referred to in this study as "Kelvin."

Because Scottish Natural Philosophy sought basic truths, it valued the application of science to determine them. Unlike Cambridge physics, which placed greater value on theory as explicated by mathematics, Scottish Natural Philosophy valued test and demonstration through experiment.

Like the string theory of atomic structure, Scottish Natural Philosophy sought to discover the underlying reason for occurrences in nature and a way to fit the natural world into one unified whole. On unity, William Hamilton, a Scottish Natural Philosopher who taught J. C. Maxwell, observed that, "It not only affords the efficient cause of philosophy, but the guiding principle of its discoveries" (as cited in Olson, 1975, p. 134).

The Cambridge wranglers—physicists who finished in the top of their class at Cambridge—commonly used analogy to explicate theory. Indeed, a good analogy was requisite to a good theory. The value of analogy is believed to have been imparted to the wranglers through the influence of Scottish Natural Philosophy. Maxwell was Scottish and attended the University of Edinburgh before Cambridge, and Kelvin was Irish and attended the University of Glasgow before Cambridge. It might be helpful to delve into the meaning of Scottish Natural Philosophy by considering the work of Dugald Stewart, one of its most noted voices, especially as a way of discerning attitudes toward metaphor and analogy.

As a Scottish Natural Philosopher, Dugald Stewart (1753-1828) has been cited as an influence on Maxwell (Harman, 1985b). Of analogy, Stewart wrote, "the principles of our nature . . . dispose us to extend our conclusions from what is familiar to what is comparatively unknown; and to reason from species to species and from individual to individual" (Stewart, 1829, p. 273). What Stewart proposes here sounds very much like Aristotle's (1952b) substitution theory in the study of metaphor: "Thus a cup (B) is in relation to Dionysus (A) what a shield (D) is to Ares (C)" (p. 693). Stewart ties analogy more closely to thought in general when he observes that we

> refer to the evidence of our conclusions, in the one case, to *experience* and in the other to *analogy*. The truth is, that the difference between the two denominations of evidence, when they are accurately analyzed, appears manifestly to be a difference, not in *kind*, but merely in *degree* [Stewart's italics] (1829, p. 273).

Here, Stewart equates analogy with thought. Interestingly enough, he seemed to anticipate postmodern objections to the idea of fixed definition.

On analogy in the sciences, Stewart (1829) credits Isaac Newton with looking for laws that would unite the heavens and Earth:

> Every subsequent step which has been gained in astronomical science has tended more and more to illustrate the sagacity of those views by which Newton was guided to this fortunate anticipation of the truth; as well as to

confirm, upon which continually grows in its magnificent conception of uniform design, which emboldened him to connect the physics of the Earth with the hitherto unexplored mysteries of the Heavens (p. 283).

This passage clearly points not only to the idea of a unification of the physical world, but specifically to the idea of seeking the same patterns in the terrestrial as in the heavens.

Kelvin and Maxwell would have been encouraged to use analogy as part of their education as scientists. Their influence in and of itself was profound upon the physicists who followed. A good example is Oliver Lodge, who was British, but an associate (and an admirer) of Maxwell. Lodge was not involved as much with the development of atomic structure (though he did transmit radio waves before Marconi), but he was Chair of the Department of Physics at Liverpool University and was extremely popular as a lecturer and as a science writer, especially writing on science for the general public, so it is appropriate to consider him as a science writer, which, as I have argued, is a type of technical communicator. Lodge was a colleague and confidant of J. J. Thomson, who is credited with first identifying the electron.

LODGE AND THE BAAS

Though Sir Oliver Lodge was not directly involved, either through studies, research, or teaching, with Cambridge, he was definitely part of its scientific cultural milieu. One of his many honors for his professional work included being elected president of the British Association for the Advancement of Science (BAAS) in 1913, but one year as president does not do justice to his life-long involvement with that group. J. J. Thomson noted that, "When Kelvin, G. F. FitzGerald, and Lodge—who was unrivaled in clear exposition—were present, one saw the BA at its best" (as cited in Jolly, 1974, p. 59). Thomson suggests here a couple of important points: Lodge's ability as a public speaker and popularizer of science, and his relationship with the BAAS.

First, the BAAS bears some explication. Founded in 1831, its primary intentions were to provide a forum for science and to serve as a conduit between science and the public, a goal now professionalized for the technical communicator. As one of the inner circle whom Morrell and Thackeray (1981) have identified as the Gentlemen of Science who founded the BAAS (p. 22), Vernon Harcourt articulated the goals of the BAAS as "'to give a stronger impulse and more systematic direction to scientific enquiry,' to promote contact between its cultivators, and to obtain greater public attention to its objects" (MacLeod, 1981, p. 17). The second and third points are important for this discussion since they emphasize the relationship the BAAS sought to develop with the public, a goal of the technical communicator. As a group of scientists with such goals, the value of scientific and technical communication becomes more apparent. MacLeod further asserts that, "For over a century, certainly until the advent of radio, the BAAS furnished the only national link between Britain's scientists and the general public" (p. 18). Unlike the Royal Society, the BAAS set out "to possess no endowment, to avoid competition with any other society, to make no collections of artifacts or instruments, to hold no property, and to let each meeting defray its own cost" (pp. 18-19). Indeed, its aims were democratic to the extent that it considered itself to be a parliament of science.

Harcourt's point of disseminating science to its "cultivators" manifested itself variously. Though one lot fell to influencing government policy, another went to education. In an 1885 address to the BAAS, the Duke of Argyll encouraged "the teaching of the young . . . not so much the mere results, as the *methods* and above all, the *history* of science" (as cited in Yeo, 1981, p. 79). As part of a proposal to bring the BAAS to Hull for a meeting, Charles Frost, the President of the Hull Literary and Philosophical Society, cited the BAAS as instrumental in "the means of developing and bringing into honorable publicity some native talent, which may have existed among us hitherto unknown or only in part appreciated, from the absence of circumstances calculated to draw it forth" (as cited in Lowe, 1981, p. 125).

The BAAS proved quite popular with a public hungry for science. Charles Lyell observed that 1,000 to 1,500 people attended the 1838 Geology section meeting alone in Newcastle (Lowe, 1981, p. 121). The BAAS also began admitting women, with 1,100 estimated to have attended the general Newcastle meeting (p. 127).

More meetings took place in centers of industry than in the shadow of the universities. Indeed, cities such as Newcastle, Hull, and Liverpool vied for the meetings, not only for the prestige but for the economic influx as well as the scientific expertise. Often, specific sections were devoted to local issues. For example, geologists with the 1846 Southampton meeting were consulted on the construction of an artesian well underway to improve the city's water supply. Facilities such as libraries and museums were often built to commemorate and accommodate a BAAS conference. Cities would also try to out-celebrate one another, even to the extent of planning fireworks. When the secretary for the Edinburgh meeting planned the fireworks, he did so with the intention that the spectacle would "beat the Cambridge fireworks hollow" (Robison, as cited in Morrell & Thackerary, 1981, p. 158).

Lodge's formal contact with the BAAS began when he was 22 and still working for his father, who sold pottery supplies. While visiting potteries in the vicinity of Glasgow and Edinburgh, he learned the BAAS would be holding a weeklong meeting in nearby Bradford (Lodge, 1932a, p. 13). The BAAS brought Lodge into contact with the Cambridge physicists, with Maxwell among them, who was one of the wranglers (Harman, 1985a).

Prior to 1873, Lodge had been aware of the BAAS. As early as the 1870 meeting, he had "read about it in the newspapers, and cut out and stuck in a press-cutting book all that I could gather of what went on there" (Lodge, 1932b, p. 16). Later, he would count Maxwell among his acquaintances, one he venerated.

Lodge's initial introduction to Maxwell occurred through what he read of him, and at the 1870 Liverpool meeting, Maxwell spoke on the relationship of physics and mathematics. He delineated how different people approach scientific knowledge. Some approach it as a mathematical expression disconnected from its existence in the physical world; others, and Lodge counted himself in this group, must have physical models for their calculations to be meaningful. "For such men," Maxwell noted, "momentum, energy, mass, are not mere abstract expressions of the results of scientific inquiry. They are words of power, which stir their souls like the memories of childhood." Maxwell was also concerned that science be popularized "and yet remain scientific." Therefore, he thought that, "For the sake of persons of these different types, scientific truth should be presented in different forms, and should be regarded as equally scientific, whether it appears in the robust form and the vivid coloring of a physical illustration, or in the tenuity and paleness of a symbolical expression" (as cited in Lodge, 1932a, pp. 20-21). Of analogy, he observes that "electrical phenomena" are being explained in terms of "dynamical phenomena." Here he speaks of the analogy drawn between mechanical processes and electricity. Of this use of language to explain science, he comments, "To apply to these the phrases of dynamics . . . is an example of a metaphor of a bold kind . . . but it is a legitimate metaphor if it conveys a true idea of the electrical relations to those who have already been trained in dynamics" (as cited in Lodge, 1932a, pp. 20-21). Lodge's early role models as scientists, then, presented metaphor and analogy as appropriate for science.

With the nature of Scottish Natural Philosophy described and its influence on Kelvin, Maxwell, and Lodge established, I now examine what role metaphor played in the development of atomic structure. Doing so is important because of how it relates to the idea of mathematics as metaphor since other scholars (Heilbron, 1985; Kuhn, 1993) have attributed the conscious decline of metaphor in science to the construction of atomic theory, when classical mechanics could no longer explain atomic structure.

THE SOLAR SYSTEM ANALOGY

The analogy of the atom as structured similarly to the solar system is probably one that more people would recognize than the explanation of the quantum atom. Consider the symbol of the Atomic Energy Commission (see Figure 1). From 1948 to 1975, the Atomic Energy Commission, a branch of the United States government, used an icon that suggested that subatomic particles circle the nucleus in regular, elliptical orbits, which describes neither the trajectory of the orbit nor the quantum leaps a subatomic particle might take from one orbit to another.

Figure 1. Seal of the Atomic Energy Commission
(Atomic Energy Commission, 2003).

This case study examines the Solar System Analogy (SSA) to atomic structure, with references to other metaphoric elements, as it manifested itself in the mid-nineteenth century, became useful as a rhetorical tool for physicists, and eventually expired as an epistemologically useful rhetorical tool. First, the extent to which the SSA is still useful today is evaluated by considering its role in educational materials for contemporary secondary students. Then, its usefulness to three pairs of nineteenth- to early twentieth-century physicists (Lord Kelvin and J. C. Maxwell; J. J. Thomson and Oliver Lodge; and Ernest Rutherford and Niels Bohr) is weighed.

Before I begin that examination, I would like to consider how Lakoff and Nunez's (2000) work frames this discussion. This chapter began with questioning mathematics as metaphorical. If we accept Lakoff and Nunez's hypothesis of mathematics as metaphorical, then the case for all language, natural and artificial, as metaphor is strengthened. In this study, the mathematics shifts abruptly with the theories of the structure of the atom. For example, Kelvin and Maxwell thought matter could not be broken down any further than the atom, so their mathematical pondering became less meaningful when the atom was found to consist of subatomic particles. Some critics of metaphor and analogy in science have blamed them for misleading science, and even proponents of metaphor such as Hesse have noted that scientists have at times centered their research around faulty metaphors that the scientists were reluctant to relinquish, but mathematics, especially in the sense that Lakoff and Nunez have argued that it has been romanticized, lends greater credence to these claims. Indeed, in Kelvin's nascent essays on atomic structure, he uses 129 equations to describe how the atom shapes matter, as illustrated in the discussion below. Which would be more credible in the opinion of most scientists seeking to add to Kelvin's work, after the atom was found to be divisible: the 129 equations or the idea of a vortex atom? The consistency lies in the use of the vortex atom, which would become the SSA, to direct the development of theory. That idea, not the mathematics describing an incorrect structure, directed future research in a more fruitful manner than the mathematics, which, as Lakoff and Nunez (2000) would probably agree, is little more than a measuring stick, though a complex and exact one. Let us now examine the more fertile SSA as it exists today.

THE SOLAR SYSTEM ANALOGY IN
SECONDARY-SCHOOL TEXTS

Today, high school chemistry and physics students are more likely than not to be exposed to the idea that the structure of the atom is analogous to the solar system. This analogy might be phrased, "As the planets orbit the sun, the subatomic particles orbit the atom's nucleus," as the simile, "An atom is like the solar system," or as the metaphor, "An atom is a miniature solar system." The CD-ROM version of *World Book Encyclopedia* tells us, "Atoms are often compared to the solar system, with the nucleus corresponding to the sun and the electrons corresponding to the planets that orbit the sun" ("Atoms," 1998), which points to the ubiquity of the SSA.

World Book, the acme of elementary-school scholarship, is not alone. A survey of secondary-school chemistry and physics textbooks by major publishers (Hewitt, 1992; Lamb, Cuevas, & Lehrman, 1989; LeMay, Robblee, & Brower, 2000; Murphy, Hollon, & Zitzewitz, 1986; Myers, Oldham, & Tocci, 2000; Stolberg & Hill, 1980; Wilbraham, Staley, Matta, & Waterman, 2000) reveals that seven out of eleven use the SSA to explain the relationship of subatomic particles to the nucleus. Of the four that do not include the SSA, three are published by Glencoe, a division of McGraw-Hill. Of these, two use the cloud metaphor (Feather, Snyder, & Hesser, 1993; Wistrom, Phillips, & Strozak, 2000), which is less accurate since the subatomic particles are in somewhat predictable orbits. Of the two that use the cloud metaphor, one tells us that, "[Niels] Bohr pictured the atom as having a central nucleus with electrons moving about it in well-defined paths" (Wistrom, Phillips, & Strozak, 2000, p. 272), which suggests the SSA. Feather et al. describe the relationship between the nucleus and the subatomic particles as being like a cloud or like a beehive, which is also less accurate for the same reason as the cloud metaphor (1993).

The third Glencoe textbook uses a chocolate chip cookie analogy and cites the work of Nagaoka (1904), whose analogy compares the nucleus and the subatomic particles to Saturn and its rings (McLaughlin & Thompson, 1999). This theory was never accorded much credibility since it suggests that the subatomic particles were in a planar orbit, but its influence over Bohr is worth examining. Though the popular pictorial depiction of the solar system suggests the planets are in planar orbit, such an assumption is not borne out empirically; their orbit is actually three-dimensional. Hence the planetary analogy is more instructive than the others cited in textbooks when it is properly modeled in the mind. In his land-mark paper "The Scattering of α and β Particles by Matter," Ernest Rutherford (1911) cites the Nagaoka model. However, though Nagaoka sought to specify the array of subatomic particles, Rutherford modeled the relationship of the atom's positive charge (Yagi, 1964). Though the Saturnian comparison in this textbook accurately reflects Nagaoka's work, in a caption to an illustration, the authors inaccurately inform us that "Nagaoka's model resembled a planet with moons orbiting in a flat plane" (Stolberg & Hill, 1980, p. 63). The fact that three of these textbooks were published by Glencoe may indicate an editorial bias against the use of the SSA.

The fourth textbook that does not include the SSA was published by Prentice Hall; however, two other Prentice Hall textbooks included the solar system analogy.[2] This fourth textbook that does not include the SSA should probably be considered the most inaccurate. First, blatant errors are forecast by poetic license invoked over J. J. Thomson's plum pudding metaphor to describe the structure of the atom. The authors describe it as "a muffin with berries scattered through it" (Frank et al., 2001, p. 78). Such a liberality with the original text could be overlooked since secondary-school students would be more likely to be familiar with berries in a muffin than with plum pudding, but unfortunately, it is accompanied by an illustration of four planets orbiting a star. Its caption reads, "His [Nagaoka's] model showed the electrons revolving around this sphere like the planets around the sun" (Frank et al., 2001, p. 78). These treat-ments ironically illustrate Thomas Kuhn's objection to the fact that "Until the very last stages in the education of a scientist, textbooks are systematically substituted for the creative scientific literature that made them possible" (1970, p. 165). While the original scientific writings would be confusing for students this age for a number of reasons, accurate depiction of those writings would be a step in the right direction.

---

[2] The other textbooks that included the solar system analogy were published by Harcourt, Brace, Jovanovich; Houghton Mifflin; Addison-Wesley; Merrill; and Holt, Rinehart, and Winston.

Historically, our conception of the structure of the atom is usually attributed to Bohr, the early twentieth-century Danish physicist, and it is frequently referred to as the "Bohr Atom." Sometimes, it is called the "Rutherford-Bohr Atom," with the addition of Rutherford because Bohr was one of Rutherford's students for post-doctoral work at Manchester, and because Bohr's work on the structure of the atom built upon Rutherford's. Rutherford hypothesized that the structure of an atom includes a nucleus and quite a bit of empty space. However, a scouring of each physicist's work reveals that neither one published a version of the SSA at key junctures in his work where it would have been appropriate.

Sometimes the SSA is attributed to J. J. Thomson (1907), who wrote, "A positively electrified ion and a corpuscle might form a system analogous to the solar system, in which the positively electrified ion, with its large mass, takes the part of the sun while the corpuscles circulate round it as planets" (p. 157). Thomson discovered the electron, the first subatomic particle to be identified through experiment in 1897. Five years before Thomson first invoked the solar system analogy, Lodge extended the analogy in what began as a series of lectures to the Institution of Electrical Engineers. These lectures were later published as "On Electrons" in the *Journal of the Institution of Electrical Engineers* and were then expanded upon in Lodge's *Electrons*, also published in 1902. Credit can certainly be allotted to J. C. Maxwell for his molecular vortex atom theory, which influenced J. J. Thomson and Lodge as well; the vortex model shows up in J. J. Thomson's (1883) work before the solar system analogy. However, it is also present in the work of William Thomson (Lord Kelvin) before it appears in Maxwell's. Though it is probably not possible to attribute the origin of the SSA to one particular scientist, this case study examines in detail the presence of this analogy in the work of these six physicists.

A NARRATIVE HISTORY OF THE
SOLAR SYSTEM ANALOGY

The idea of the SSA can certainly be traced into the past far preceding Kelvin. Kuhn (1957) paralleled twentieth-century scientific revolutions initiated by Bohr and other physicists with the Copernican. More importantly for the SSA, Kuhn includes the ancient Greek atomists Leucippus (480-420 BCE—dates approximate) and Democritus (460-370 BCE), who theorized "an infinite universe containing many moving earths and many suns" (1957, p. 236). Similar to the Scottish Natural Philosophers in their intent to find an underlying design, Leucippus and Democritus sought to merge the structure of the cosmos with the structure of the atomic world. For the ancient Greek atomists, there must be particles and a void to allow movement. The atom, however, was indivisible for the ancient Greeks, and while an infinite number of atoms moving into an infinite void suggests more about cosmology than about atomic structure, the use of the astronomical to describe the atomic (and vice versa) is worth noting. The idea of a vortex to describe atomic motion also appears in Democritus' and Leucippus' atomic theories (Pullman, 1998). This examination of the vortex metaphor begins with Kelvin, but he credits Hermann von Helmholtz with his inspiration for this metaphor. That this study does not further trace the metaphor points to science as a socially constructed act. It would easily be possible to trace the SSA's antecedents even further. For example, on Oliver Lodge's use of the SSA, Rowland (1990) asserts that Lodge was inspired to use it from exposure to Balfour Stewart's and P. G. Tait's *The Unseen Universe* (1876). Rowland also attributes the SSA to Thomas Young. Furthermore, Kuhn (1957) has delineated how René Descartes sought to explain the movement of atoms in the void. For Descartes, atoms move through the void in what Kuhn refers to as "circulatory streams" (p. 240), which eventually form into vortices that become solar systems.

LORD KELVIN

Considering Kelvin's molecular vortex theory as a precursor to the solar system analogy is like thinking about the solar system itself forming from spinning clouds of gas. The impetus for this inquiry arose from his work with electricity and magnetism. In essence, what Kelvin sought to understand was the nature of magnetism and electricity. Did these forces have a physical structure? Scientific analogies were drawn with gravity, but unlike gravity, magnets attract and repel as well as polarize. Another theory proposed that magnets exude a type of magnetized matter. Kelvin, however, began to equate magnetism with electricity. Maxwell's interpretation of Kelvin led Maxwell to equate magnetism, electricity, and light in his electromagnetic equation, which resulted in a series of equations still standard in physics and engineering.

Kelvin and Maxwell enjoyed a mentor-student relationship. Maxwell credited Kelvin with the idea of vortex molecules from his 1856 paper, "Dynamical Illustrations of the Magnetic and the Helicoidal Rotatory Effects of Transparent Bodies on Polarized Light." At that point, Kelvin was wondering whether the structure of electricity and magnetism were "finite vortical or other relative motions of contiguous parts of a body; it is impossible to decide, and perhaps in vain to speculate, in the present state of science" (as cited in Sharlin, 1979, p. 122). This passage is important because it illustrates the nascence of molecular vortex motion. To be "vortical" is to have a vertex, or center point, for the structure of the molecule. The point was to discover the molecular structure of forces such as magnetism, electricity, and light.

Kelvin's article under consideration is titled "Hydrodynamics," and it actually consists of two articles. The first is titled "On vortex atoms" and was first published as part of the *Proceedings of the Royal Society of Edinburgh* and reprinted in 1867 in *Philosophical Magazine*, then the leading publication for physics research in England. The second one is titled "On vortex motion" and was published as part of the *Transactions of the Royal Society of Edinburgh* in 1869. The text notes it was first read "29th April, 1867."

Kelvin (1910) begins by noting the work of German physicist Hermann von Helmholtz on vortex motion, which was derived from the concept of vortex motion in a perfect liquid. Kelvin believed that Helmholtz's model represented actual atomic structure. P. G. Tait, another British physicist, is credited with contributing to Kelvin's conception of the vortex atom. For the purposes of demonstration, Tait devised a box with a diaphragm. Smoke was created in the box (sometimes by simply blowing in cigar smoke), and then it was expelled by tapping on the diaphragm. For Kelvin, the resulting smoke rings represented a model for the building blocks for all matter, which would be found in motion. This structure of matter also attempted to account for a unifying common matter that Kelvin and other physicists called the aether, which would fill in not only the void of outer space but the void between molecules and even between atoms. Kelvin's vortex theory accounts for the motion of atoms in the aether. Molecularly, these atoms are the ancient Greek's, moving into the void: "simple bodies [that] should have one or more fundamental periods of vibration, as has a stringed instrument of one or more strings, or an elastic solid consisting of one or more tuning-forks rigidly connected" (p. 3). It is interesting to note the allusions here to musical imagery in the shape of a "stringed instrument" and a "tuning fork." Music as an analogy is appropriate because it is, in a sense, composed of particles whose motion creates a coherent whole. Alluding to Lucretius, Kelvin concludes "the molecule of sodium, for instance, should not be an atom, but a group of atoms with void space between them" (p. 3). Kelvin's models, then, were influenced by Helmholtz's concept of vortex motion in a liquid, so from the very beginning, the SSA is portrayed as a social construction since, though I use Kelvin's work as a starting point, his theory was informed by Helmholtz and Tait, and to some extent, by the ancient Greeks. In addition, Tait's model added to his conception of a model born of motion. All of these are important factors as the SSA begins to coalesce.

The problem, according to Kelvin (1910), is accounting for the strength of the vortex atom. How can it be strong and composed of empty space? The aether could not in itself be strong enough, so the answer for Kelvin lies in creating some type of structure. This structure must be created, then, by particles in rapid motion, which he imagined as "a finite mass of incomprehensible frictionless fluid completely enclosed in a rigid fixed boundary" (p. 13), a spherical structure and within it, "infinitesimal circular rings" (p. 19). Two diagrams Kelvin included in his paper illustrate the structure of matter. The first illustrates his idea of molecular structure (see Figure 2). These rings are Kelvin's idea of molecular structure. Within the rings, Kelvin thought the atoms to be arranged in concentric circles (see Figure 3). Kelvin drew this final illustration from Maxwell's published work, so Maxwell's work next becomes the focus.

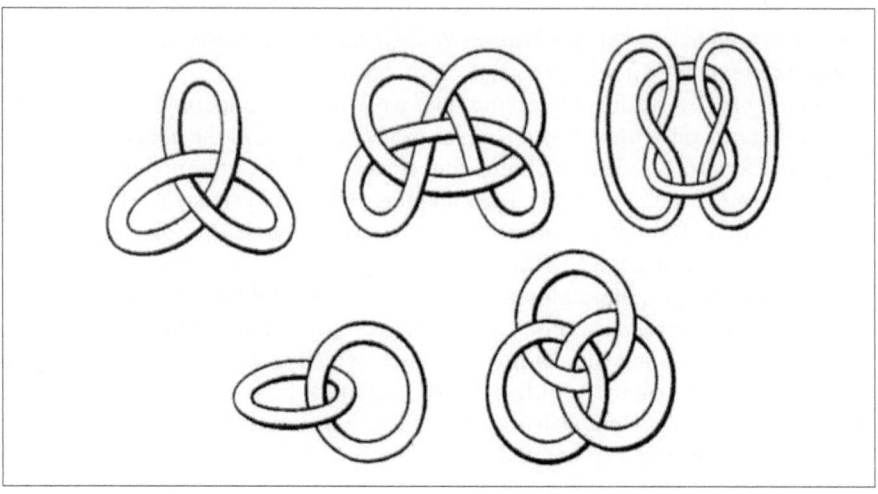

Figure 2. Kelvin's vortex rings.

Kelvin (1910) begins to give the SSA shape, literally, as spherical and with concentric rings depending upon a center, though he does not presume to propose what that center may consist of, if it consists of anything. His antecedents lie not only with Helmholtz but with Lucretius. His explanation, though abstract, is one that is beginning to take shape. Its fertility becomes apparent as the study progresses. The vortex atom was useful to Kelvin because it allowed him to assign a shape to the structure of electricity. Maxwell takes the vortex theory a step further.

JAMES CLERK MAXWELL

The way Maxwell (1986a, 1986b) recognized the unity of electric- and magnetic-field structure is regarded as one of his most significant contributions to physics (the other is his work on gases). The problem he addressed concerned how light can travel through space, which was believed to be composed of aether, a gas similar to air, though lighter. Maxwell hypothesized that there must be some similarity between light and magnetism, which would also align these with electricity, because they would need to be structured similarly at the molecular level for them to pass through the aether. Therefore, Maxwell proposed that all of these functioned like a system of spinning cogwheels, or vortices, at the molecular level (Goldman, 1983).

Maxwell (1986b) introduced astronomical models into his vortex theory with a penny as a model. Gravity on such an object does not vary according to its weight or shape. The penny weighs the same whether it is balanced on its edge or lies flat.

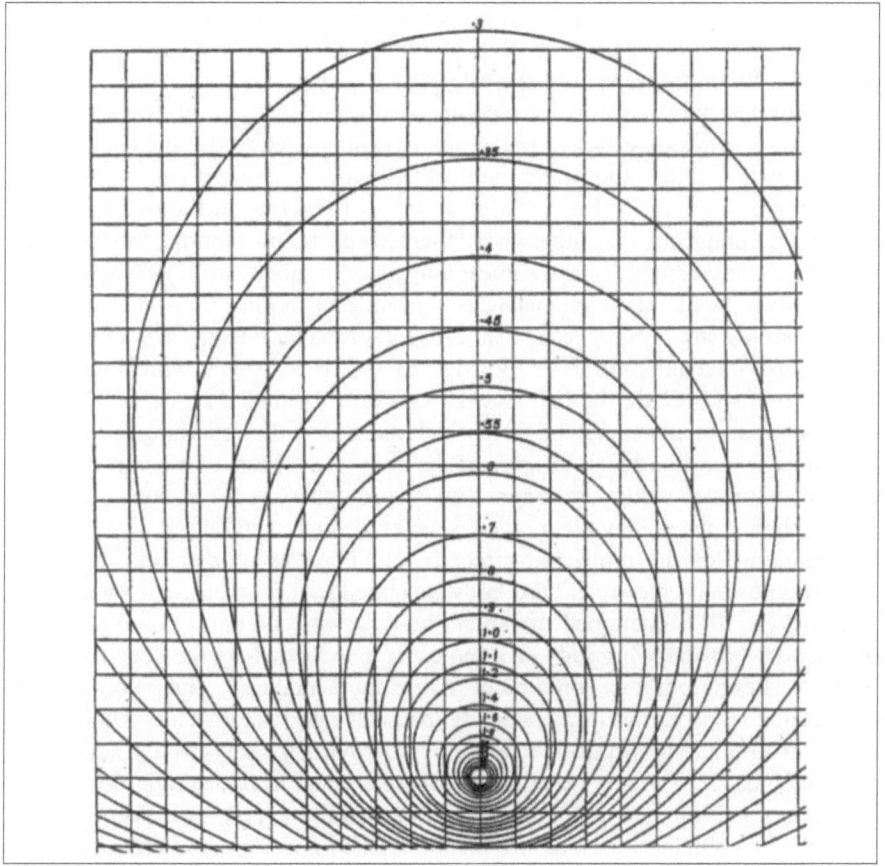

Figure 3. Maxwell's vortex rings (Thomson, 1910, p. 63).

Maxwell maintains that these facts prove the penny is not solid because if it were, then whether it lies on its side or edge would cause it to differ in its gravitational attraction, much as a sail differs by what direction it takes to the wind. With these concepts thus hypothesized, he moves more directly to the astronomical by noting that "If it were not so then the sun and earth together would attract the moon less during an eclipse of the moon than at a full moon"; in other words, the moon's orbit would be affected by its relative position to the sun and Earth (Maxwell, 1986b, p. 172). Such astronomically-inspired analogies are indicative of a direction for Maxwell's thought that again manifests itself as the SSA.

An interesting place to further examine Maxwell's explication of the atom is in an 1875 *Encyclopedia Britannica* article. Because an encyclopedia is created for the general public, such a publication is worthy of consideration for this study.

First, it is worth noting that Maxwell defines an atom as "a body which cannot be cut in two" (1986a, p. 176). Both he and Kelvin believed the atom to be indivisible. When they speak of the role of empty space and the aether, they are thinking of it as between atoms that compose molecules. Maxwell then compares and contrasts previous views of the atom, harkening back to the ancient Greeks, who also believed the atom to be indivisible, before moving to current opinion. For example, similarly to Kelvin, Maxwell cites Lucretius on the role of empty space by pointing out that otherwise, "there could be no motion, for the atom which gives way first must have some empty place to move into" (1986a, p. 177). On the other hand, Maxwell balances this idea by noting that "the opposite school maintained . . . every part of space is full of matter, that there is a universal plenum, and that all motion is like that of a fish in water, which yields in front of the fish because the fish leaves room for it behind" (1986a, p. 178). As a contemporary touchstone, he notes Helmholtz's work regarding a vortex's ubiquitous features and Kelvin's use of the theory as a basis of his vortex atom. Maxwell moves toward astronomical metaphors when he poses molecules "as elastic spheres" when they are merged, such as through metallurgy.

Accounting for the molecular structure of light was problematic. Maxwell (1986a) notes that "the small hard body imagined by Lucretius, and adopted by Newton," (p. 201) functioned to explain material objects, but it could not offer explanation "for the vibrations of a molecule as revealed by the spectroscope" (p. 202). These vibrations could be described mathematically, but Maxwell saw that such an accounting was actually bad science, because the rigidity of molecular structure was conceived as a way of providing a basis for the solidity of material objects. To simply account for them mathematically is to ignore the fact that their vibration contradicts this idea of rigidity. Therefore, according to Maxwell, "To obtain vibrations, we must imagine molecules consisting of many such centers being separated" (p. 202). It is important to note this distinction because it bears on the way theory is developing toward the Solar System Analogy. Maxwell's theory could account for the data, so

> the disciple of Lucretius may cut and carve his solid atoms in the hope of getting them to combine into worlds; the follower of Boscovich may imagine new laws of force to meet the requirements of each new phenomenon; but he who dares to plant his feet in the path opened up by Helmholtz and Thomson [Kelvin] has no such resources. His primitive fluid has no other properties than inertia, invariable density, and perfect mobility, and the method by which the motion of this fluid is to be traced is pure mathematical analysis. The difficulties of this method are enormous, but the glory of surmounting them would be unique (p. 203).

The "mathematical analysis" Maxwell refers to is that which is derived from classical mechanics, another way in which the solar system analogy begins to emerge. Of course, the idea that "the motion of this fluid . . . is pure mathematical

Analysis" becomes more important to Bohr and the solar system as a fertile metaphor. It is interesting to note that Maxwell chides "the disciple of Lucretius" for attempting to shape "his solid atoms in the hope of getting them to combine into worlds," which also seems to foreshadow the use and dispensation of the SSA. Indeed, the SSA would shape atomic structure theory in the sense of the type of analogy drawn between the subatomic particles as worlds unto themselves. Bohr (1913) would ultimately resort to the mathematical metaphor to describe the atom. Maxwell would answer that the mathematical suggests a rigidity that was not borne from his conception of the atom, but Bohr certainly did struggle long with the SSA.

Maxwell questions the extent to which the law of gravity has been considered in terms of its relation to atoms. As a result, he is drawn to consider the solar system itself as a model:

> Now, we know that the effect of the attraction of the sun and the earth on the moon is appreciably different when the moon is eclipsed than on other occasions when a full moon occurs without an eclipse. This shows that the number of corpuscles which are stopped by bodies of the size and mass of the earth, and even the sun, is very small compared with the number which pass straight through the earth or the sun without striking a single molecule. To the streams of corpuscles, the earth and the sun are mere systems of atoms scattered in space, which present far more openings than obstacles to their rectilinear flight (1986a, p. 205).

If the sun and Earth do not stop all corpuscles, then how is a ball falling to Earth accounted for? According to Maxwell, the corpuscles in a ball must be moving at a much greater velocity.

Maxwell notes that the light measured by a spectroscope, which breaks light up into its constituent bands, suggests that the light of stars, nebulai, galaxies, and comets are composed of the same material (1986a, p. 208). Therefore, he reasons, if hydrogen is believed to be the same, whether it is burned in the sun, on Arcturus, or on Earth, then hydrogen's molecular structure can be assumed to be same when the hydrogen is derived from Earthly elements (1986a, p. 210).

The SSA, then, is beginning to emerge. Maxwell brings the subject of gravity to the arena of discussion, as well as other astronomical entities. He validates mathematical expression, but without eliminating models. He finds the vortex model useful because it explained how electricity, magnetism, and light could travel through the aether. And he added the idea of motion to his model, which became more useful for explaining the structure of the subatomic realm.

J. J. THOMSON

One of the most important scientists to pick up the idea of vortex rings was J. J. Thomson, who is remembered today principally for his discovery of the electron. He also contributed significantly to delineating the structure of the atom.

The first clear evidence of Thomson (1883) delving into atomic and molecular structure appears in *A Treatise on the Motion of Vortex Rings*, which won Cambridge University's 1882 Adams Prize and was later published by Macmillan and Company while Thomson was a mathematics lecturer with Trinity College. Thomson has attributed his interest in vortex rings directly to Kelvin (Davis & Falconer, 1997). At this point, Thomson's contribution seems to be the way he attempted to relate vortex rings to periodic law, which groups elements according to their atomic numbers. As a model, Thomson directs his readers to think of the vortex rings as similar to a screw joined round to its head. Each thread "will represent the central line of the vortex core" (p. 119), which suggests that at this point, Thomson was beginning to think of the atom as "being built up of subatomic particles" (Davis & Falconer, 1997, p. 16). These may be joined into chains that form molecules, according to Thomson, an idea reminiscent of Kelvin's vortex rings. To understand molecular structure, Thomson proposes "systems of vortices placed at the angular points of the polygon as the primaries and the component vortex rings of the primaries as the secondaries of the system" (pp. 118-119). However, the SSA does not yet emerge in his work. Though there is no central point of orientation for the molecules, the SSA could have unified theory, especially from a classical mechanics perspective. Instead, molecules were believed to be encased in the jelly of the aether. By 1904 Thomson described the structure of an atom as a sphere

> of uniform positive electrification, and inside this sphere a number of corpuscles arranged in a series of parallel rings, the number of corpuscles in a ring varying from ring to ring: each corpuscle is traveling at a high speed round the circumference of the ring in which it is situated, and the rings are so arranged that those which contain a large number of corpuscles are near the surface of the sphere, while those in which there are a smaller number of corpuscles are more in the inside (pp. 254-255).

Though the SSA is suggested by this description, it is not yet named. Indeed, Thomson mixes his metaphors by the end of his paper when he writes, "We have taken the case of the four corpuscles as the type of a system which, like a top, requires for its stability a certain amount of rotation" (p. 265). Mixing metaphors does not significantly create a problem with clarity and is typical rather than atypical of scientific metaphors. Furthermore, such an approach is healthy in a sense because a mixed metaphor is one to which the scientist is less tied, allowing for easy dispensation when the metaphor ceases to be useful. Such an observation suggests the trajectory of shape a metaphor can take as it develops in a scientist's work.

On mixed metaphor, what Lakoff and Johnson (1980) have observed is especially pertinent for science: "There is no one metaphor that will do. Each one gives a certain comprehension of one aspect of the concept and hides the others. To operate only in terms of a consistent set of metaphors is to hide many aspects of reality" (p. 221). Lakoff and Johnson intend their comment for language use in general, but for the scientist, shifting theories demand greater metaphorical acuity. Such was especially the case when scientists were depending to such a great degree upon instrumentation for observation.

In Thomson's work on the cathode ray tube, the SSA begins to be articulated. First, it is worth noting that Thomson announced his theory of the electron, the first subatomic particle to be identified, in 1897, from his experiments with the cathode ray tube. At that point, Thomson had already begun to work with Ernest Rutherford, who arrived at the Cavendish laboratory as one of Thomson's students.

A cathode ray is produced when electricity passes through a vacuum tube and strikes phosphorescence. The presence of an electron explains why the cathode ray can pass through foil, which is composed of atoms. How could an object composed of atoms pass through another object composed of atoms, especially since they were considered the building blocks of matter? The answer had to be that there were subatomic particles. Hence, what were passing through the foil were not whole atoms themselves but subatomic particles. Thomson identified the electron as a subatomic particle.

The term "electron" did not immediately emerge in Thomson's work, and his deduction of the existence of subatomic particles did not meet with immediate agreement from other physicists. Thomson himself did not begin calling them electrons, a term that was in use since it had been coined by Irish physicist G. J. Stoney (1891), who theorized the electron. Thomson, on the other hand, referred to the subatomic particles he believed to have identified as "corpuscles" (Davis & Falconer, 1997, p. 123).

Continuing his work on subatomic particles, Thomson's 1903 article "The Magnetic Properties of Systems of Corpuscles Describing Circular Orbits" uses astronomical metaphors to describe the movement of particles "rotating uniformly in a circular orbit" (p. 681). That he uses the word "orbit" indicates he is moving toward using the SSA, but so does the fact that these particles are "rotating" since the planets in the solar system rotate as well as orbit. However, Thomson was not entirely certain about the complexity of the structure of the atom. He also proposes to

> suppose the atoms of a substance, like the atoms of radio-active substances, were continually emitting corpuscles; the velocity of projection of the corpuscles under consideration being, however, insufficient to carry them clear of the atom, so that the corpuscles describe orbits round the centre of the atom: then if, the motion of the corpuscles were not accompanied by dissipation of energy, the corpuscles would not endow the body with either magnetic or diamagnetic properties; if, however, the energy of the corpuscles dissipated during their motion outside the atom, so that they ultimately fell with but little energy into the atom, a system consisting of such atoms would be paramagnetic. If the energy of projection were derived from the internal energy of the atom, there would thus be a continual transference of energy from the atom to the surrounding systems . . . (p. 689).

In this passage, Thomson is ruminating on the complexity of the structure of the atom. What role, he wonders, will magnetism, or the lack of a magnetic field, play upon atomic structure? If magnetism can alter a flow of atoms, what is the effect on the structure? At this point, the complexity of the structure seems to demand too much for the SSA to be useful. As a matter of fact, the problems that he addresses here are the same ones that would cause Niels Bohr to reject the SSA in favor of a mathematical model. If the "corpuscles" do not radiate energy, then they would fall "into the atom," as Thomson phrased it, since the nucleus itself had not yet been proposed. At this point, Thomson is trying to account for the presence of the electron in terms of its relationship to the atom, which was not divisible at this point. He is supposing that the atom is emitting the "corpuscles," what he would come to call electrons. Still, the SSA is taking shape as Thomson begins to use astronomical terms such as "orbit" to arrange the atom with a center around which his corpuscles move.

A year later, in 1904, Thomson wrote an article titled, "On the Structure of the Atom: An Investigation of the Stability and Periods of Oscillations of a Number of Corpuscles Arranged at Equal Intervals around the Circumference of a Circle; with Application of the Results to the Theory of Atomic Structure." He begins with,

> The view that the atoms of the elements consist of a number of negatively electrified corpuscles enclosed in a sphere of uniform positive electrification, suggests, among other interesting mathematical problems, the one discussed in this paper, that of the motion of a ring of n negatively electrified particles placed inside a uniformly electrified sphere (p. 237).

Such an arrival at a theory of "corpuscles" in a sphere occurred because there were two theories of atomic structure: one that supposed the subatomic elements were stationary and the other that theorized that they were in motion. The idea of motion was favored because stationary particles, though they would lend to the structure greater stability, could not account for what would happen when electrons were added. Motion allows for adding electrons, but the question remains over what holds the particles in a state of attraction. Thomson answers by placing his corpuscles in a shell of positive electricity (Rayleigh, 1943).

Since the corpuscles were now in orbit within this shell, Thomson (1904) drops the astronomical imagery. "Orbits" become "rings" as metaphorically, though certainly not literally, he recalls vortex ring imagery. Indeed, the idea of vortex rings was that the rings were molecules and the atoms were arranged inside them. As I shall show, Bohr, too, shifted from "orbit" to "ring" as a metaphor.

Next, Thomson mathematically works through the effect of the displacement of the corpuscles according to their number in a system. He also describes the arrangement of magnets in the Mayer experiment as an analogy (1904, pp. 238-255).

With these results in mind, Thomson (1904) proceeds to consider the structure of the atom. In particular, he is interested in where the particles would be and how their placement would affect the atom's properties. He proposes that the results of the study at that point indicate that "the corpuscles will arrange themselves in a series of concentric rings" (p. 255), which implies the SSA, but the idea of these particles being in orbits suggests gravity, and the way in which the movement of these corpuscles could be bent by magnetism indicated that gravity would be too strong a force. Furthermore, though a large number of corpuscles arranged on one plane would be unstable, they could be stabilized if they were not planar, in which case they would be in concentric shells. Of this idea, Thomson admits that he has not worked out all the complexity of the schematics. He proposes that this type of atomic structure would make similar use of what chemists make of atomic weight. However, since he cannot hope to work through the problem in that article, he continues to calculate the structure of the atom according to the number of rings. As a result, he returns to elements of the SSA. When he analyzes corpuscular vibration, he notes the corpuscles must be divided into two sets: "Those arising from the rotation of the corpuscles around their orbits" and "Those arising from the displacement of the ring from its circular figure" (p. 259). It is only at this point that he returns to a structure resembling the SSA. When he begins to discuss what could happen when subatomic particles are added or subtracted from a system, he reverts to speaking of rings, perhaps because of the role gravity would play in a solar system when a member is added or subtracted.

Thomson (1904) concludes by accounting for a radioactive atom, which, as he knew, emitted particles. The radioactive particles, according to Thomson, are circling the center more slowly, which makes them less stable. In this case, the SSA may have been helpful since a weaker gravitational field would make the atom less stable, but Thomson opts instead for, "we have taken the case of the four corpuscles as the type of system which, like a top, requires for its stability a certain amount of rotation" (p. 265). The "like a top" simile suggests gravity, only on a smaller scale than that of planets. Also, a top emits only energy by its motion, an aspect that becomes important in Bohr's work. Another of Thomson's points is that the element with particles orbiting more slowly would also decay more slowly as well. Comparing the structure of radioactive atoms to a top can be read as an example of meiosis, or understatement. Perhaps Thomson decided not to use the SSA because he did not want to deal with the gravity issue.

A year later, in 1905, a paper delivered to the Royal Institution of Great Britain finds Thomson more favorably disposed to the SSA. In the first paragraph, he summarizes his work with the cathode ray tube and with subsequent suggestions of subatomic particles emanating from "incandescent metals, from metals illuminated by ultra-violet light, and radio-active substances" (1997, p. 217). Thomson begins to engage in analogy as he considers the fact that negatively charged "corpuscles" will repel. Supposing an atom to be a group of corpuscles, Thomson recognizes that something must cause them to adhere. If, then, "the corpuscles form the bricks of the structure, we require mortar to hold them together" (p. 217). He concludes that "positive electricity acts as the mortar," to hold the corpuscles together (p. 217).

Thomson continues by citing Kelvin on the idea of the atom as a sphere containing positive electricity and corpuscles. He depicts graphically a geometric relationship of the corpuscles, moving from a single corpuscle to an "octahedron with 2 inside" (p. 218). If there is a larger number of corpuscles, Thomson notes, predicting their positions is difficult. At this point, he cites Mayer's magnetized needles stuck through corks. As they are tossed into a bowl of water, they arrange themselves in geometric shapes; three will be a triangle; four will form a square, and five, a pentagon; but with the sixth needle, one will shift to the center with the other five continuing to hold in a pentagon. With a seventh needle, the others will form a circle around the one in the middle. He notes, of course, that the model is planar, unlike the actual structure of an atom. With this idea in mind, Thomson returns to atomic weight as a way of determining atomic structure. Because Thomson (1997) thought he was dealing with the building blocks of matter, he proposes that the difference between one element and another is the "rearrangement of the positive electricity and corpuscles" (pp. 223-224).

Thomson notes that he has calculated the potential energy of an atom according to the number of corpuscles. He explains his findings in this regard as

> analogous to the case of a number of stones scattered over a hilly country . . . the stones, if subject to disturbances, would run from the hills into the valleys, and though the stones might be uniformly distributed to begin with, yet in the course of time, they would accumulate in the valleys (p. 223).

This statement is accompanied by a simple line graph. The corpuscles with more potential energy would be on the peaks and would be less stable, while the corpuscles with less potential energy would be more stable and tend to accumulate in the valleys.

When Thomson (1997) turns his attention to "*Chemical Combination. Action of the Atoms on each other*," he again approaches the Solar System Analogy:

> As far as I know, the only cases in which the conditions for equilibrium or stable steady motion of several bodies acting upon each other have been investigated, is that suggested by the solar system; the case in which a number of bodies—suns, planets, satellites—attract each other with forces inversely proportional to the square of the distance between them (p. 224).

The fact that Thomson begins with, "As far as I know," suggests that he is not completely satisfied with the SSA. However, it is remarkable that he begins to formulate it here, since Rutherford would not begin to hypothesize the nucleus for another five years, and Rutherford's article on the topic would not be published until 1911.

Thomson (1997) continues with "The complete solution of this problem, or anything approaching a complete solution, has proved to be beyond the powers of our mathematical analysis" (p. 224). He then continues by summarizing some of Maxwell's work with the stability of Saturn's satellites and concludes that this body of work indicates that a large, central mass must be apparent to maintain the satellites' stability, and more than six satellites "saturates" the planet with satellites. Thomson, at this point, seems to accept a limit on the number of satellites (corpuscles) and that the motion of the corpuscles creates magnetic attraction. Therefore, the magnetic and positive electricity provide the atom with stability. As a result, he was willing to accept the SSA.

Two years later, in 1907, in his book *The Corpuscular Theory of Matter*, Thomson felt more strongly about making a commitment to the SSA: "A positively electrified ion and a corpuscle might form a system analogous to the solar system, in which the positively electrified ion, with its large mass, takes the part of the sun while the corpuscles circulate round it as planets" (p. 157). Perhaps the shift from journal article to book suggested a wider readership, and he felt the SSA would be more valuable as a communication tool. In this case, the SSA would be generative for his audience.

Though the SSA is more commonly referred to as the "Bohr Atom" or the "Rutherford-Bohr Atom," it is evident that Thomson did a great deal of early work on atomic structure. His metaphors varied somewhat, ranging from a top, to scattered stones, to the SSA. In his work is a movement toward the SSA, but it is not arrived at quickly or easily, perhaps because, unlike the top or the stones scattered in the valley, he saw that other scientists would invest to a great degree in such a scientific metaphor. The top and stones seemed tossed off, more similar to modifying metaphors that serve as a brief springboard of thought rather than a serious model than can be extended and refined.

Thomson is also interesting as a transition figure in the development of the SSA since he begins with the vortex model. Through his work, the SSA developed at a time when scientific discoveries, most notably his hypothesizing of the electron, began to shape the metaphor. With Thomson, the SSA grew and bloomed, which points to its fertility and its usefulness as a rhetorical tool for the scientist.

We can see today that Thomson erred on a number of issues, but the process of discovery is fraught with falsification, as Popper (1972) noted. Thomson contributed quite a bit to early studies of the structure of the atom to such an extent that England became where Bohr wanted to study the atom. However, one of Thomson's colleagues, Oliver Lodge, felt even more at home with the SSA. He wrote an entire book on it well after Bohr had dispensed with it.

OLIVER LODGE

Oliver Lodge was one of Thomson's colleagues and peers, and his work was doubtlessly influenced by Maxwell's analogical approach, as I have presented it, and along with G. F. FitzGerald and Oliver Heaviside, science historian B. J. Hunt (1991) refers to Lodge as one of the Maxwellians in his book of the same title. According to Hunt, these Maxwellians "transformed the rich but confusing raw material of the [Maxwell's] *Treatise* [*on Electricity and Magnetism*] into a solid, concise, and well-confirmed theory . . . the 'Maxwell's theory' we know today" (p. 2). Lodge not only followed Maxwell's lead as he studied the subatomic realm, but his PhD focused on electricity, and he also used astronomical analogies to describe the subatomic.

Lodge is somewhat different from other physicists because his work did not contribute directly to the development of the theory of atomic structure. Though he was a professor of physics at University College, Liverpool, where he chaired the department of physics for many years, as a popularizer of science, he contributed to theorizing the structure of the atom. His influence did not extend only to the general public; his work, in this instance through his public lectures, influenced Japanese physicist Hanataro Nagaoka (1904), whose theory of atomic structure influenced Bohr's metaphorical approach.

The first recorded instance of Lodge's explication of the Solar System Analogy occurs in his 72-page 1902 essay, "On Electrons," which represents a lecture presented to the Institution of Electrical Engineers.

˙ Throughout this essay, the SSA is gradually taking shape in a more mature, realized fashion as he begins building the suggestion of the astronomical: "The mobility of diffusiveness of a gas depends on its mean free path, and that depends on its atomic size; the smaller it is, the more readily it can escape collision. Hence it is that collisions are so rare in astronomy: the bodies are small compared with the spaces between them" (1902, p. 64). On the speed of the subatomic particles, he further extends the analogy by noting that "No known speed which can be conferred upon matter is sufficient to bring this latter effect into prominence. The quickest available carriage is the earth in its journey around the sun . . . " (1902, p. 67). Again, as with Thomson, it is remarkable that he would use this analogy, since Rutherford would not hypothesize the nucleus for eight more years.

Just prior to his clearest exhibition of the solar system analogy, Lodge introduces an idea he explores later in more detail in his 1924 *Atoms and Rays: An Introduction to Modern Views on Atomic Structure and Radiation:* that of "atomic astronomy, with atoms and electrons instead of planets and satellites" (1902, p. 85). With this idea, he means to unite the universe in the conception of matter. For example, he compares the action of the electrons in the cathode ray tube to the Aurora Borealis "on a gigantic scale" where "the earth's magnetic lines of force are illuminated by flying electrons from the sun entangled and guided by them" (1902, p. 88). He thought "that before long evidence will be forthcoming on this and other lines, which will enable us to accept or reject the hypothesis of the electric nature and unification of matter" (1902, p. 102).

The idea of atomic astronomy bears further explication. For example, at the end of the fifth chapter of *Atoms and Rays*, Lodge tells us, "Gradually we are beginning to understand more and more about the mechanism of this marvelous universe; and it is instructive to find the same law and order ruling everywhere—inside the atom and in the remotest depth of space" (1924, p. 63).

In "On Electrons," Lodge arrives at his solar system analogy: "Even inside an atom of mercury, therefore, the amount of crowding is fairly analogous to that of the planets in the solar system. For though the outer planets are spaced further apart than the inner ones, they are also bigger, to practically a compensating extent" (1902, p. 98). The last sentence is probably the most important since it suggests that there is quite a bit of empty space in an atom, about nine years before Rutherford (1911) was able to theorize this idea. Such a hypothesis illustrates the fertility of the Solar System Analogy.

As an extension of this analogy, Lodge (1902) entertains the idea of comets as they might be compared to what happens when one metal is fused with another. Just as when one metal fuses to another, a comet might join a solar system, eventually dissolving into meteors.

Five years later, in his book *Electrons or the Nature and Properties of Negative Electricity* (1907), Lodge is beginning to doubt the SSA. It falls last as "A fifth view of the atom would regard it as a central 'sun' of the extremely concentrated positive electricity at the center, with a multitude of electrons revolving in astronomical orbits, like asteroids, within its range of attraction" (p. 150). The use of "asteroids" instead of planets is interesting because it suggests a somewhat less lifelike entity than a planet, which radiates energy, unlike asteroids that are today recognized as a potential danger in the solar system, since a stray one could collide with the earth.

Of the other four models, he lists the first one as the atom consisting of "ordinary matter . . . associated with sufficient positive electricity . . . to neutralise the charge belonging to the electron." This atom balances the then-known subatomic particle, the electron, against the rest of the atom. The second one proposes that "the atom may consist of a multitude of positive and negative electrons, interleaved . . . and holding themselves together in mutual attractions. . . ." This model is more like what we believe about the atom today. Since no data at this point indicated the atom consisted of other parts, it must be listed second because it can be only a supposition. Lodge is perhaps reasoning here that if there is one subatomic part, then there may be others. The third supposes the atom might be "an indivisible unit of positive electricity, constituting a presumably spherical mass or 'jelly' . . . " (1907, p. 148). The idea of "jelly" reprises the role of the aether as that which filled in all voids, whether between atoms or between the stars in outer space. The fourth is an "interlocked admixture of positive and negative electricity, indivisible and inseparable into units, and incapable of being appreciably sheared by applied forces" (p. 149). The next-to-the-last one, then is the indivisible Greek atom that does not account for the electron unless it is an emanation from the atom.

While such dissatisfaction is healthy, it is worth noting that Lodge became restless with this analogy about six years before Bohr (1913) laid it to rest. The problem with the SSA is that according to classical dynamics, such a system would have to radiate energy continuously or it would eventually collapse, with the subatomic particles falling into the nucleus since they do not radiate energy as the planets do. In this sense, they would be more like meteors than comets, much less planets. However, his interest revived by the time he wrote *Atoms and Rays* in 1924, because he evidently held quantum mechanics suspect, and perhaps because he thought the SSA could communicate more clearly to a general audience.

By the 1920s the idea of quantum mechanics had caused the scientific world to reject the Solar System Analogy, since it does not work well as a model when electrons jump from one orbit to another, or as Lodge would have it, "modified . . . by that at present mysterious limitation, or condition 'the quantum' about whose real meaning we are still in the dark," which suggests that Lodge did not entirely believe that quantum mechanics answered problems posed by the relationship of subatomic particles. Of course, neither did Albert Einstein. Lodge concludes with, "It must suffice to say that we are living in the dawn of a kind of atomic astronomy which looks as if it were going to do for Chemistry what Newton did for the solar system" (1924, p. 203).

That a physicist would extend such an analogy to the extent that it permeates a book is remarkable in light of our current expectations about scientific writing. However, he by no means limited himself to astronomical imagery. In "On Electrons," he variously referred to electrons as "cannon-balls" (1907, p. 48) and "bullets" (p. 60). Electricity can be conducted, according to Lodge, by the "bird-seed method," the "bullet method," or the "fire-bucket method," depending upon the medium of conduction (p. 79). Furthermore, hyperbole is invoked since, according to Lodge, "matter, moving with the speed of light, would have enough energy to lift the British Navy to the top of Ben Nevis" (1907, p. 51). These different tropes are a sign of a healthy approach to metaphor as a part of the generative process.

Lodge is important for his role as a popularizer of science, and the SSA was valuable for him for that purpose. He influenced Nagaoka (1904), who is discussed in more detail in his relation to Rutherford's (1911) work, because Rutherford cites him at the end of his article that establishes the nucleus as the center of the atom. Yagi (1964) has pointed out that Nagaoka attended at least one of Lodge's public lectures in London. However, Lodge's influence on Thomson should not be discounted. They were friends and colleagues who not only fraternized through the BAAS, but dined in one another's homes (Jolly, 1974). The SSA was bound to have been bandied between them.

ERNEST RUTHERFORD

To contribute to the theory of the atom's structure, Ernest Rutherford suggested the idea of the nucleus and empty space in an atom, as opposed to the aether. Though the SSA is often referred to as the Rutherford-Bohr Atom, he did not make as much use of it as Thomson or Lodge or much movement toward it, neither in his essay that theorized the nucleus, nor in a later article he wrote for the more general readership of *Scientia*.

As the first 1851 Exhibition Science Scholarship student in physics from the University of New Zealand, Rutherford began studying atomic structure when he arrived in 1895 at the Cavendish Laboratory to work with J. J. Thomson. Though his initial interest was in wireless transmission, he quickly became interested in atomic structure after Rontgen's 1895 discovery of X-rays. At a February 13, 1896, London Royal Society meeting, J. J. Thomson, Rutherford's mentor, reported, "The passage of these rays through a substance seems thus to be accompanied by a splitting up of its molecules, which enables electricity to pass through it by a process resembling that by which a current passes through an electrolyte" (as cited in Feather, 1973, p. 40). Such a research atmosphere certainly influenced Rutherford's work as well; he continued to study atomic structure for the rest of his life.

What is regarded as Rutherford's most important contribution to atomic structure is his idea of the atom having a very small nucleus, where the focus of the atom's charge may be found. Experiments indicated that alpha particles from radioactive substances were affected by the angle of their deflection when they passed through a thin mica sheet. When alpha particles were directed toward gold foil, not only were more deflected, but some were stopped completely, as if, Rutherford commented, "one had fired a large naval shell at a piece of tissue paper and it had bounced back" (as cited in Campbell, 2001), a comment that indicates some propensity for metaphor. As an analogy, however, the SSA is not apparent.

Hypothesizing the atom's nucleus to serve as its center in terms of mass and electrical charge that held the atom together allowed Rutherford to account for the deflections and the reversal of the alpha particles. Otherwise, the deflection of its particles would disintegrate the atom. Such a proposition also suggested that a good deal of empty space accounts for the rest of the atom.

In considering Rutherford's work, two articles are examined: one from *Philosophical Magazine* (1911) that presents his proposed nucleus for an audience of peers and the other from *Scientia* (1963), an article written for a more general audience. The *Scientia* article was originally published in 1914 and is significant because it serves as a segue to Bohr's work. The *Philosophical Magazine* article, "The Scattering of α and β Particles by Matter and the Structure of the Atom," begins with the deflections of α and β particles as premises. Rutherford then states, "it has generally been supposed that the scattering of a pencil of α or β in passing through a thin plate of matter is the result of a multitude of small scatterings by the atoms of matter traversed" (p. 669). The "pencil of α or β" is a metaphor, but it is descriptive rather than epistemological, more like a literary metaphor than a scientific one. He then notes that Geiger and Marsden's work indicates that the material the particles are aimed at has something to do with the amount and type of deflection. By the end of the paragraph, he concludes "A simple calculation shows that the atom must be a seat of an intense electric field to produce such a large deflection" (p. 669). To place the atom in the center of such a field, the SSA seems an obvious choice, especially since Rutherford was intimately familiar with J. J. Thomson's work, which further examination bears out.

Addressing Thomson's work on the atom, Rutherford points out that to balance corpuscles of positive and negative energy at the center of the atom depends on small deflections. When these deflections increase in magnitude (or do not occur), then it must be because of the influence of a central charge. Rutherford concludes that

> it seems simplest to suppose that the atom contains a central charge distributed through a very small volume, and that the large single deflexions are due to the central charge as a whole, not to its constituents. At the same time, the experimental evidence is not precise enough to negative the possibility that a small fraction of the positive charge may be carried by satellites extending some distance from the centre of the atom (p. 687).

Here Rutherford proposes the idea of a central point with "satellites." The idea of satellites suggests entities that are in orbit, but there is no direct comparison between the atom and the solar system. There is also the notion that the nucleus serves as an anchor for the atom, much as the sun in the solar system. In addition, Rutherford has not ruled out here that these satellites carry "a small fraction of the positive charge." The SSA ultimately fails as analogy, as do many science analogies because one item is not the other. The subatomic particles do not seem to do more than *bear* a charge, unlike the solar system's planets that *radiate* the energy that prevents them from being drawn into the sun. Experimentally, Rutherford had not been able to prove that the subatomic particles do not radiate energy. As a matter of fact, this article more resembles a thought experiment in that it does not report empirical research; rather, it speculates upon the work of others, most notably that of J. J. Thomson, Geiger and Marsden (1909), and Geiger (1910).

Rutherford (1911) may not have chosen to use the SSA because he was convinced that the source of the attraction between the subatomic elements was electrical, not gravitational. Instead of the SSA, he uses the much more pedestrian "pencil" metaphor several times (pp. 669, 673, 680-683), and this to describe the shape of α or β rays, not an individual atom or how it is structured. At the end of the article, he refers to the subatomic particles as satellites (p. 687), and admits that any conclusions drawn from experiment cannot deny that the satellites may bear enough of a charge to prevent them from falling into the nucleus, but again, he fails to invoke the SSA. The answer to this careful rhetorical posing lies with Rutherford's theory that the electrical holds the atom together, as his repeated use of the metaphor "charge" suggests. "Charge" is certainly a metaphor in this case since the force holding the atom together was still unknown and a matter of debate and study. Jeanne Fahnestock (1999) has noted how the use of words like "charge" and "current" indicates an eighteenth-century debate over whether electricity is more like water or firearms.

In the conclusion to this article, Rutherford discounts another celestial analogy, that posed by Japanese physicist Hanataro Nagaoka, whose Saturnian atomic model predates Rutherford's. In 1904, Nagaoka drew an analogy with Maxwell's work on the constancy of Saturn's rings with atomic structure:

> The system I am going to discuss, consists of a large number of particles of equal mass arranged in a circle at equal angular intervals and repelling each other with forces inversely proportional to the square of distance; at the centre of the circle, place a particle of large mass attracting the other particles according to the same law of force. If these repelling particles be revolving with nearly the same velocity about the attracting centre, the system will generally remain stable, for small disturbances, provided the attracting forces be sufficiently great. The system differs from the Saturnian system conceived by Maxwell in having repelling particles instead of attracting satellites. The present case will evidently be approximately realized if we replace these satellites by negative electrons and the attracting centre by a positively charged particle. The investigations on cathode rays and radioactivity have shown that such a system is conceivable as an ideal atom. In his lectures on electrons, Sir Oliver Lodge calls attention to a Saturnian system which probably will be of the same arrangement as that above spoken of (pp. 445-446).

Nagaoka's references are to Lodge's lectures. Nagaoka studied in Munich and Berlin, and he visited England. While Rutherford (1911) recognizes Nagaoka's work (1904), he also dismisses it because the Saturnian atom sought to offer a schematic to locate the subatomic particles, but Rutherford's more delicate tweaking sought a pattern of the atom's positive charge (Yagi, 1964). Rutherford also responded cordially to a letter from Nagaoka and recognized that Nagaoka would "notice that the structure assumed in my atom is somewhat similar to that suggested by you in a paper some years ago. I have not yet looked up your paper; but I remember that you did write on the subject" (as cited in Yagi, 1964, p. 39). In his article "The Scattering of α and β Particles by Matter and the Structure of the Atom," Rutherford (1911) writes of Nagaoka's Saturnian structure that, "from the point of view considered in this paper, the chance of large deflexion would practically be unaltered, whether the atom is considered to be a disk or a sphere" (p. 688). In essence, Rutherford is saying that the overall shape of the atom does not matter in terms of the attraction between the nucleus and the subatomic elements, but his rejection of an astronomical analogy is not as readily dismissible. Certainly there is a difference between electricity and gravity, especially since electricity was in the process of becoming harnessed. It is curious that Rutherford would slight the SSA, especially when his central point of attraction could be equated to the sun. The solar system itself, after all, is not planar but three dimensional.

It is worth noting that while the idea of the nucleus is considered Rutherford's important contribution to the study of the structure of the atom, another one that sprang from this study is the idea that the atom consists of quite a bit of empty space, because the nucleus was conceived as being much smaller than the atom itself, and the electrons were proportionally smaller and proportionally farther away. Rutherford imagined this empty space as being occupied by an electric field created through the interplay of the positively and negatively charged particles, an idea left over, perhaps, from that of the aether pervading space. Again, though the idea of empty space and there being quite a bit of it between the nucleus and the subatomic particles suggests an analogy that the SSA would lend itself to, the attraction between the nucleus and the subatomic particles seemed to Rutherford to be electrical rather than gravitational. Still, the orbit metaphor continued to be useful to describe the movement of the subatomic particles.

"The Structure of the Atom," Rutherford's *Scientia* article (1963) written for a more general audience in 1914, finds him searching for a metaphor as he deftly avoids the SSA. He first surveys discoveries that led to the structure of the atom, beginning with the discovery of Röntgen rays in 1895. One problem he addresses is the lack of proof of positive subatomic particles, except in matter in general. He describes the structure of an atom as "like the coats of an onion" (p. 449). As a metaphor, this comparison is problematic since it suggests that the subatomic particles are a bit too stable. It does not allow for electrons in motion. Rutherford follows this metaphor by noting that, "Sir J. J. Thomson has examined mathematically in detail the possible stable distributions of electrons in one plane and has deduced the possible arrangements of the electrons for a number of different values" (p. 449). However, his next move is quite significant: "The Thomson atom has undoubtedly served a very useful purpose in giving a simple and easily understood idea of atomic structure" (p. 449). Rutherford does not specify, though, what he means by the "Thomson atom," other than the onion metaphor he mentions. Thomson does use a number of images (the SSA, a top, plum pudding), but he had more recently mentioned the SSA. By using an onion, Rutherford seems to be setting up a straw man. Such an easily discounted metaphor could make room for a new one while still containing some descriptive value.

Rutherford (1963) continues with, "It has the great advantage that the law of force involved admits readily of mathematical calculation and the position and number of the electrons in different rings can be directly calculated." Layers of onions do not readily admit to mathematical calculation, not in the way of planetary orbits, which are suggested by the description of "electrons in different rings." The telling is in Rutherford's conclusion to this paragraph:

> The ultimate test of any atomic model lies, however, in its ability to explain the experimental facts, and we shall see reasons for believing that this type of atom must be much modified before it is capable of accounting for any unexpected phenomena which has been brought to light in recent years (p. 449).

The "unexpected phenomena" in this case were the variance of the deflections of particles passing through foils. How could Thomson's cathode ray pass through foil? An onion's layers do not suggest subatomic particles dynamic enough. Indeed, it is more conceivable that a solar system could lose a planet, or even more so, that a sun could lose a comet. Given the number of different types of subatomic particles now known to compose the atom, and the different types of relationships, it is not surprising that the SSA survives in secondary-school textbooks. But at the time, Rutherford's concern was to promote electrical charge as the unifying force, rather than gravity.

Rutherford (1963) concludes the article by citing the problems with describing the atom in terms of the relative positions of the electrons. Such a discussion is important because the positions can neutralize the atom's charge. Furthermore, he notes problems created by electrons that radiate energy. Eventually, these particles would expend their energy and fall into the nucleus. He then refers to Bohr's work as specifying that "the radiation of energy from an atom can only take place in certain definite ways," and he notes Bohr's adoption of Planck's quantum theory, though he also suggests that Bohr's work is by no means complete.

Rutherford clearly deviates from the SSA. While it would be especially valuable for explaining the structure of the atom, he consciously avoids it, which was probably because of his devotion to the idea of electricity, rather than gravity, as the force that held the atom together. It is worth noting that Thomson also thought a positive charge, not gravity, held the atom together, yet he used the SSA to a greater degree than Rutherford, and he certainly worked fruitfully in the study of the atom's structure. His work is especially noteworthy in terms of the ways in which the SSA contributed to his experimental work and offered interpretation of the data derived from experiment.

Though Rutherford does not invoke the SSA, it casts a long shadow over his work. For example, one of his objections to atomic structure was that the electrons would crash into the nucleus if they were not continuously radiating energy. In this case, the SSA could have been instructive because, as theory suggests, an analogy can lead us to construct a theory or to falsify it. Through the process of what Popper (1972) would call falsification, the SSA could have been valuable. If it could not show what an atom is, then the SSA could indicate what an atom is not. In that sense, it could be fertile.

NIELS BOHR

Niels Bohr became a central figure in the study of the atom and atomic energy in the twentieth century. The SSA is typically attributed to him; he made use of it when communicating to general scientific audiences, such as in his 1922 Nobel Prize acceptance speech, or lectures to lay audiences. However, he is frequently cited as the one who laid the SSA to rest (Heilbron, 1985; Kuhn, 1993).

Hailing from Denmark, Bohr came to Cambridge in the fall of 1911 for postdoctoral study with J. J. Thomson. He had just finished his doctorate in physics from the University of Copenhagen, and his dissertation focused on the problem of the mechanics of electrons in metals, specifically under what conditions electron collisions occur when a metal is exposed to electricity. He decided that there must be a good deal of space between atoms, but while some calculations supported his hypothesis, it could not support why metals were stable solids.

Cambridge did not work out for Bohr as well as he would have liked. He was able to work there, but he did not feel his labors were particularly pro-ductive. Perhaps just as important, he was not able to foster a relationship with Thomson.

Part of the problem with Thomson may have occurred with their first meeting, when Bohr critiqued some of Thomson's work on the atomic structure of metals. Specifically, Bohr pointed out errors with what J. L. Heilbron (1985) has called "The Electron Theory of Everything," which means that many physicists, including Thomson, thought that the material world could be explained as being composed of clusters of electrons (p. 34). Such a supposition preceded Rutherford's theory of the nuclear atom, and it depended upon work with metals that hypothesized that "electrons move through metals as do ions through a dilute solution or molecules through a perfect gas" (Heilbron, 1985, p. 34). The goal, again, was to find laws governing electrons and molecules that were applicable to all substances. However, heat and electricity did not seem to be accounted for by this theory. H. A. Lorentz argued in 1905 that he had quantitatively amended any differences for gases (Heilbron, 1985). Bohr's doctoral dissertation countered that Lorentz's theory could be successfully subjected to experiment if it were assumed that "the mean free path [would] change with velocity, an assumption he [Lorentz] thought equivalent to introducing a force between electrons and metal molecules" (Heilbron, 1985, p. 35). Therefore, according to Bohr, electrons can interact with molecules other than upon collision, which is important because

Lorentz's theory did not account for heat radiation and magnetism, areas where these ideas did not hold up to experiment. By 1903 Lorentz addressed this issue with a new theory "applicable to radiation whose vibration period is much longer than the average time interval between successive collisions of an electron with metal molecules" (Heilbron, 1985, p. 35) that seemed to parallel Max Planck's radiation formula. However, Planck's formula deviated widely from Drude and Lorentz's hypothesis, and Thomson thought he had reconciled those differences. Unfortunately, Bohr's extensive calculations indicated that, according to Heilbron, "Thomson's program was hopeless" (1985, p. 35).

Bohr thought Thomson would appreciate being informed of these mistakes. Bohr also presented Thomson his recently completed dissertation and asked him to read it, hoping he might help him to publish it in England. There is no evidence that Thomson ever read it, but there is no evidence that his lack of response evidenced any ill-will toward Bohr. Many have attributed it to Thomson's notorious absentmindedness (Moore, 1966, pp. 32-33).

Though Bohr worked in Thomson's Cavendish laboratory, he very quickly came under the influence of Rutherford, who visited the lab in December 1911 and lectured on his recent work on the discovery of the atom's nucleus. The meeting began a lifelong friendship, and in 1912 Bohr left Cambridge for Rutherford's lab in Manchester.

Bohr's time in Manchester was certainly fertile, so fertile, in fact, that at times, he alone is credited with the SSA; there is an argument for that because he wrote a great deal about the structure of the atom. In a trilogy of articles published in *Philosophical Magazine* in 1913, he wrote about 78 pages that attempted to explicate what he referred to as "Rutherford's model." These articles are regarded as contributing to the understanding of the atom in the following ways:

- For any conceivable movement of electrons around the nucleus, only a few specified orbits are allowed, and those follow specific parameters;
- Though classical mechanics suggest that orbiting electrons should emanate energy, electrons are prevented from doing so;
- Leaping from orbit to orbit, electrons produce energy necessary to maintain the system (Gamow, 1964, p. 52).

Though these ideas revolutionized physics in more ways than one, some unfortunate fallout is the abandonment and discrediting of models such as the SSA as a part of the scientist's epistemology.

In the 1913 trilogy, Bohr makes no direct reference that compares the atom to the solar system. Instead, he is much more careful than that. Indeed, he begins the first article by attributing to Rutherford's theory the idea that "the atoms consist of a positively charged nucleus surrounded by a system of electrons kept together by attractive forces from the nucleus" (1913, p. 1). Throughout his three articles, Bohr returns to these ideas, telling us, for example, that, "General evidence indicates that an atom of hydrogen consists simply of a single electron rotating round a positive nucleus of charge e" (p. 8). In general, his attitude suggests that he is very conscious of having absorbed this idea from Rutherford. However, he seems reluctant to embrace it completely, perhaps because Rutherford, who at that point was his mentor, did not.

Many have felt that Bohr was not comfortable with the idea of regular orbits, and later research bore him out since it is now believed that the subatomic particles are in predictable, though not necessarily regular, orbits around the nucleus. Gravity cannot explain the orbit because the particles would need to generate energy to avoid being pulled into the nucleus. Indeed, the electrons do not seem to do more than bear a negative charge.

The notion that Bohr (1913) abandoned the SSA is somewhat accurate, but to say that he abandoned metaphor would be mistaken. Instead, he shifts to a new metaphor, that of a ring. Indeed, this shift becomes more evident as the series of articles progresses. Table 2 illustrates this shift.

Bohr begins using the ring metaphor in Part II, "Systems Containing Only a Single Nucleus," to the greatest extent where he discusses the configuration of the electrons' orbit in relation to the nucleus. He saw this orbit as planar. Nagaoka's (1904) work might have contributed to this idea since Rutherford (1911) discusses Nagaoka's theory, though as I have noted, he dismisses it. Bohr was doubtlessly familiar with Nagaoka's theory because Rutherford cites it in his article that established the nucleus as the center of the atom. Bohr also corresponded with Nagaoka. In Bohr's collected papers (1981), a postcard from Nagaoka dated December 27, 1913 reads, "Hearty thanks for your kindness in sending me several papers on atomic structure; it seems to be intimately connected with the Saturnian atom, with which I was occupied about ten years ago" (p. 549). The exact circumstances for the postcard are unclear because it is not known whether Bohr sought Nagaoka's input on his theory. The brief postcard does not suggest it, and the three articles were published in July,

Table 2. Distribution of "Orbit" and "Ring" Metaphors in the "On the Constitution of Atoms and Molecules" Trilogy (Bohr, 1913)

	Orbit	Ring
Part I	16	21
Part II	20	148
Part III	14	72
Totals	50	241

September, and November of 1913. Perhaps Bohr does not cite Nagaoka's work because Rutherford dismissed it, and Bohr was working at that point in Rutherford's laboratory and under his tutelage to some degree. Additionally, Nagaoka saw the structure of the atom as literally planar, like Saturn's rings. In a departure from traditional mechanics, Bohr saw the atom as did Rutherford, as a three-dimensional entity with electrons whose planar orbits were projected at angles that would have broken the Saturnian model. From the quantum mechanics model, the jump of one electron to a different orbit as a way of maintaining the energy of the orbit created problems for the SSA as well.

Bohr (1913) did maintain the "orbit" metaphor to the very end of the final article of the trilogy. As to why he mixed his metaphors and did not make a metaphorical commitment can be explained in more detail by examining the idea of metaphor from a cultural perspective.

Considering his educational background offers further illumination. In the introduction to the problem of his dissertation, Bohr (1981) tells us that, "Lorentz's theory is based on the following mechanical picture" (p. 298). Some sort of metaphor or analogy could be reasonably expected to follow. Instead, Bohr proceeds with what he would have believed to be the ostensible:

> In the interior of metals both atoms and free electrons are assumed to be present. The dimensions of the atoms and the electrons, i.e., the ranges within which they affect each other appreciably, are assumed to be very small compared to their average mutual distances; thus they are thought to interact only in separate collisions, in which they behave as hard elastic spheres. Moreover, the dimensions and masses of electrons are thought to be so much smaller than those of atoms that collisions among the free electrons can be neglected compared to collisions between electrons and atoms, and that atoms can be regarded as immovable in comparison to the electrons (Bohr, 1981, Vol. 1, pp. 298-299).

The point here is not that Bohr should have used the SSA or necessarily any metaphor. He is less likely to have used the SSA since he completed his dissertation on April 12, 1911, the same year Rutherford published "The Scattering of α and β Particles by Matter," so Bohr did not learn of Rutherford's conception of the nucleus until he had arrived in Cambridge for postdoctoral study. In Bohr's "Electron Theory of Metals" (his dissertation), there are also other instances of the use of metaphor, such as "path" to describe the movement of electrons, but at an instance when Bohr seems to promise a metaphor, he does not deliver it, which points to the lack of facility with metaphor exhibited in the trilogy. Further examination of his dissertation reveals that Bohr does not use any metaphors beyond those endemic to electricity, such as "current" or "path" to describe the route of electrons. The atoms themselves are referred to as "spheres," either "hard" or "elastic." Bohr's diagram (Figure 4) of the relationship of the electrons

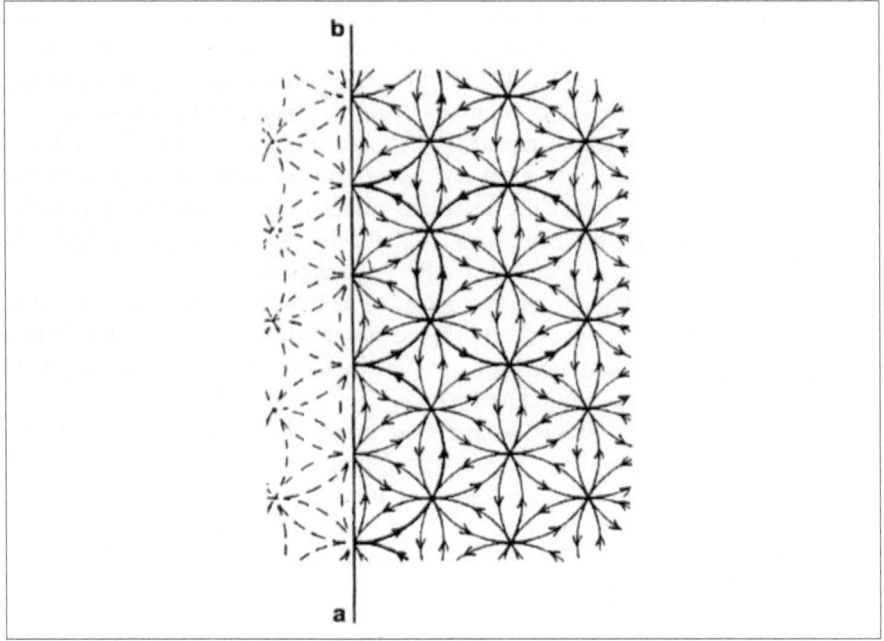

Figure 4. From Bohr's "Electrons and Atoms" (1981, p. 381).

to the atom reveals a firmer grasp of a visualization of the atom and the electrons than his verbal description suggests.

Leon Rosenfeld, who edited Bohr's collected works, refers to this depiction as a "regular lattice" (1981, p. ix). However, the point of this study is not for rhetoricians to dictate how science should be communicated, but to report how it has been communicated and to open (or repave) new avenues of communication for scientists to pursue. Bohr could have used a metaphor here, but he chose not to. Why? The answer may lie in his education.

Niels Bohr's doctoral dissertation is interesting to examine because a doctoral dissertation is a reflection of the pedagogy that shapes a student. In Bohr's case, his father's influence is relevant, because Christian Bohr was a physiologist and a professor at the University of Copenhagen where his son matriculated. One of the elder Bohr's friends, Christian Christiansen, a physics professor at the University of Copenhagen and a family friend, introduced Niels to physics as an undergraduate and guided him through his doctoral dissertation. Niels and his brother Harald, a gifted mathematician, were "admitted as silent listeners to philosophical conversations of his father and his friends" (Rosenfeld, 1981, p. xx). In a Steno lecture to the Danish Medical Society in Copenhagen in 1949, Niels notes his father's influence.

The title of the lecture is "Physical Science and the Problem of Life," and he discusses the problems encountered with studying organic life at the atomic level. In the lecture, Bohr cites what Rosenberg refers to as "a concise statement" of C. Bohr and his colleague's response to their generation of professors' "reaction against the mechanistic materialism of the preceding generation" (1981, p. xx). The "mechanistic materialism," especially in light of C. Bohr's comments on analogy, can be read as relating to the role of analogy in the sciences and offer some explanation for how and why Bohr used analogy as he did. On analogy, C. Bohr commented,

> By means of analogies which so easily present themselves among the variety of organic functions, it is merely the next step to interpret this functioning from a subjective judgment about the special character of purposiveness in the given case. It is evident, however, how often, with our so narrowly limited knowledge about the organism, such a personal judgment may be erroneous, as is illustrated by many examples (as cited in N. Bohr, 1958, p. 96).

That C. Bohr would suggest that analogies are useful as a tool that should be used carefully is evident in the way that N. Bohr used metaphors and analogies, for he did not lavish them upon his work. He was more likely to use a metaphor like "path," "current," or "orbit" than to commit to an analogy such as the SSA, and it is questionable to what extent N. Bohr, especially the young N. Bohr, recognized "current" as a metaphor. His father concludes with these thoughts: "But one thing is what may be conveniently used by the preliminary investigation, another what justifiably can be considered an actually achieved result" (as cited in N. Bohr, 1958, p. 96), which suggests that an analogy, though part of the thinking process, should be ultimately dismissed. It is also worth noting that C. Bohr's article was first published in Danish and later translated into English by N. Bohr himself.

CONCLUSION

One of the most important links with the theoretical literature presented in Chapter Three is with what McMullan (1976) refers to as *U*-fertility (unproven fertility) and *P*-fertility (proven fertility), or what I have referred to as the metaphor or analogy's "generative" quality. The SSA is certainly fertile in the sense that it provides a gateway into subatomic physics for students. As Petrie and Oshlag (1993) point out, what may be a dead metaphor for the teacher can be interactive for the student. In terms of its fertility, it is important to note that both Thomson and Lodge were writing of the SSA about eight years before Rutherford, one of Thomson's students, began thinking about the nucleus. As Thomson and Lodge first discussed the SSA, it definitely exhibited *U*-fertility. To some degree, Rutherford picked up the analogy, though he does not name it directly in his article that hypothesizes the nucleus. Still, he used terms such as "orbit" to describe the movement of the electron around the nucleus, and the SSA is often referred to as the Rutherford-Bohr atom. Rutherford's use of the unnamed metaphor is a pivotal moment because he both utilizes and dispenses with it, passing it off to Bohr, who treats it in a similar manner. For Rutherford, the SSA became Aristotle's "nameless act," the sowing around the god-created flame (1952b, p. 693), and my interpretation has argued that Bohr does not so much dispense with the SSA as commit a paradigm shift as he reverts to ring imagery that can be thought of as influenced by Nagaoka as well as Maxwell and Kelvin, and indirectly by Lodge.

On the generative quality of metaphor and analogy, I have noted why Thomson and Lodge became restless with the SSA. Still, for both scientists, the SSA extended their theories epistemologically beyond what was known at the time, and correctly so. Both Thomson and Lodge anticipated that the atom had a center and that there was quite a bit of empty space between the subatomic particles years before Rutherford theorized both of these ideas. With his onion metaphor, Rutherford seemed to be setting up a straw man. Such a use of metaphor shows, if anything, the way Rutherford discounted metaphor; it seems to serve as an easy substitution for the SSA and is positioned there to support Rutherford's idea that electricity, not gravity, permeated the SSA. Though an onion does not suggest an electrical field, neither does it suggest gravity, which was evidently more important for Rutherford. Regardless, the SSA can be read as forming his theory through falsification.

Bohr's decision to drop the SSA can be argued as cultural. Kelvin and J. C. Maxwell emerged from the Scottish universities where metaphor and analogy were valued perhaps even more than among Cambridge's wranglers. Thomson, too, was British, as well as a second wrangler, and he valued metaphor as a model. Though Lodge was neither Scottish nor Irish, he was Presbyterian, and the Presbyterian Church is, to this day, the state church of Scotland, and his epistemological values are consistent with Scottish Natural Philosophy. On the other hand, Rutherford was a New Zealander, and though he was under the sway of Thomson as a research student, he had completed his undergraduate degree in New Zealand. Niels Bohr completed his doctorate in Copenhagen before he came to study with Rutherford. Though metaphor can in no way be read, even in the sciences, as exclusively an icon of Scottish universities or of Cambridge since metaphor transcends language in the sense that all natural languages are expressed through metaphors, it was valued there in a way that it may not have been valued as a tool to the scientist outside of those cultural environs. One way, then, in which metaphor's fall from grace in the sciences can be read is as a cultural schism. The SSA began to break down when it passed from Thomson to Rutherford, at least partially because Rutherford and Bohr's educational background did not value metaphor to the same extent as in the British Isles. It might be argued as well that Thomson did not make metaphor the center of his epistemological process to the same extent as Kelvin, Maxwell, and Lodge, with their stronger connections to Scottish Natural Philosophy.

On the mathematical as metaphorical, I have pointed out that while Kelvin and Maxwell's mathematics failed, the vortex metaphor survived and eventually became the SSA. Indeed, Maxwell considered describing atomic vibrations mathematically as "bad science" because the rigidity of the molecule's structure accounted for the solidity of objects. Maxwell certainly valued mathematics, but not at the expense of models or good science.

Certainly mathematics was important to these other scientists as well, as they sought to describe the atom quantitatively. But the fact that none of them were completely right about the atom devalues the mathematical calculations and points to them as metaphorical. Certainly each was able to make the mathematics work to some degree, so the mathematics created another model that was nonetheless an incorrect one, which means that the mathematics was, on the one hand, as equally whimsical as the metaphors, and on the other hand, just as epistemological because the mathematics sought to represent the models, when they were present. The symbolic value of mathematics is certainly to some degree rhetorical: not everyone understands it, so it functions as a code for the initiated. When the uninitiated encounter it in a scientific article, the article's claims have greater ethos.

Kuhn has criticized the educational process of scientists because, "until the very last stages in the education of a scientist, textbooks are systematically substituted for the creative scientific literature that made them possible" (1970, p. 165). It is difficult to influence how scientists are taught in science classes, but in technical communication courses, analogical thinking can be taught. Students can be exposed to it when definition is taught, it can be praised when it is used at least competently, and students can be assigned to write analogies. As more and more students are required to take writing courses beyond the core curriculum, such an opportunity becomes clearer.

With the idea of the mathematical as metaphorical in these studies established, we can now move on to the idea of natural language as metaphorical by examining a more contemporary instance of metaphor usage, this time in the life sciences. Such a shift in the sciences will further broaden this study's claims for metaphor in science.

CHAPTER 5

The Question of Metaphor
in Natural Language:
A Case Study

This chapter continues the examination begun in Chapter 3 of whether or not a natural language contains words that are metaphoric and words that are not. While theorists such as I. A. Richards (1936), Max Black (1962), and Chaim Perelman and Lucie Olbrechts-Tyteca (1969) have supported the idea that all language is metaphoric, they all seem to assume that there is a literal and a metaphoric meaning. How can all language be metaphoric, yet contain words with both a literal and a metaphoric meaning?

Aristotle is one source of this idea. He does not commit to the idea that all language (or all thought, as Nietzsche would have it) is metaphoric.When Aristotle tells us that that "the greatest thing is to be a master of metaphor. It is the one thing that cannot be learnt from others; and it is also a sign of genius, since a good metaphor implies an intuitive perception of the similarity in dissimilars," he suggests that metaphor is a special gift (1952b, p. 694). If it is a special gift, then it is something that not everyone has access to or can be successful with. Indeed, metaphor as a sign of genius alludes to the idea of the "born writer," that writing is something a student either can or cannot do, a misunderstanding that is the bane of the writing instructor. This idea is further reinforced in our culture when we learn of Samuel Taylor Coleridge waking from an opium dream to write "Kubla Khan," a standard offering even in secondary-school textbooks, and then being unable later to finish the poem as he had originally envisioned it. Aristotle (1991) further tells us that effective use of metaphor cannot be learned from others. Metaphor, it would seem, is a great gift that not all have access to as they have access to language, until one considers the metaphors that primitive cultures without a written language invent to describe the creation of the world and other phenomena they otherwise find inexplicable, which suggests that, to the contrary, metaphor is universal.

To universalize metaphor further, it should be remembered that Aristotle (1991) ties metaphor to enthymeme, which links it to rhetoric, and the ability to persuade is certainly a very basic human need. While to participate fully in Athenian culture one needed to be an able speaker, it can be argued that adept argument is a life skill, as the persistence of public speaking and writing courses in the college curriculum demonstrates to some extent.

Though these considerations are worthwhile to present a balanced representation of Aristotle's metaphor, it is important to remember that although Aristotle made no blanket pronouncements on metaphor and language, he placed metaphor at the seat of created language. Boyd (1993, p. 484) argued that Anglo-Saxon words may have at one time "cut the world at the joints," which offers some support for the idea of a language that is not metaphoric. Unfortunately, we do not possess a time machine that would allow us to witness the creation of language. And even if we did, where would we begin, with guttural sounds applied to objects? Where would we stop in a quest for the source of language, to determine which word is metaphorical and which is not? Aristotle's example of the "god-created flame" (1952b, p. 693) is evidence enough of the way that humans have invented language since ancient times, and further speculation is not likely to be fruitful. What is fruitful is to recall the way *energia* is exemplified in the god-created flame. It is with *energia* that the new meaning occurs, and for that reason, Aristotle thought metaphor to be worthy of philosophy.

On the subject of a literal language, it is clear that many rhetoricians are more similar to the author of the *Rhetorica ad Herennium* than to Aristotle in the sense that the author of the *Rhetorica ad Herennium* attributes to metaphor little more than the quality of decoration. In a sense, I. A. Richards (1936) falls into that camp. On the one hand, Richards claims for metaphor that "all thought is metaphoric" (1936, p. 94), "we cannot get through three sentences of ordinary discourse without it [metaphor]," and "the pretense to do without metaphor is never more than a bluff waiting to be called" (p. 92). Such statements are Aristotelian and Nietzschean. However, he allows that, "Even in the rigid language of the settled sciences we do not eliminate or prevent it [metaphor] without great difficulty," which actually suggests less metaphoric usage in language than limiting it to Anglo-Saxon words that "cut the world at the joints," (Boyd, 1993, p. 484) or by some other criteria that will turn metaphor into "discussable science." Furthermore, he would have us believe that words can be "simultaneously literal and metaphoric," which suggests Berggren's (1962/1963) idea of stereoscopic vision, but the stereoscopic approach also supports the idea of a literal use that is not metaphoric. In Richards' defense, he recognizes that metaphor creates language, that what we think of as literal language is "super-imposed upon a perceived world which is in itself a product of earlier or unwitting metaphor." For that reason, Richards concludes that "if we take the theory of metaphor further than it was taken in the eighteenth century, we must have some general theorem of meaning" (1936, pp. 108-109). What little theory he proposes is inconsistent and inadequate, however.

Perelman and Olbrechts-Tyteca (1969) are similar to Richards (1936), though they have explored metaphor more thoroughly. Epistemologically, they value metaphor, but at some point it must be dismantled, like scaffolding, when the metaphor has become a scientific law. Such a transition suggests that there is a point when a literal language takes over to express the scientific metaphor. For example, with "light is a wave" or "light is a particle," there is a point at which these characteristics of light can be demonstrated, either literally, with a prism, or through Descartes' hypothesis that light is contained in a medium, which his readers interpreted with the wave metaphor. More recently, Perelman (1982) credited analogy only with "affirming a weak resemblance and . . . useful for formulating hypotheses, but . . . [to] be eliminated in the formulations of the results of scientific research" (p. 114), which again suggests that metaphors must be discarded in favor of a literal interpretation. Indeed, while noting the generative value of scientific thought, Perelman (1989) values the mathematical over the metaphorical because mathematics is an "artificial language," without metaphor, that "the natural languages, with their lesser development, should strive to imitate" (Perelman & Olbrechts-Tyteca, 1969, p. 130). The previous chapter's discussion of mathematics as metaphoric stands in opposition to this idea, however.

Perelman and Olbrechts-Tyteca (1969) further align with Richards (1936), and Black (1962), for that matter, in how they name the parts of the metaphor. What Richards calls tenor and vehicle, Perelman and Olbrechts-Tyteca name theme and phoros. Black calls these parts the frame and the focus. What the naming of parts indicates is that there is a part of the metaphor that is metaphoric and a part that is not. It is in the non-metaphoric part where literal meaning is thought to reside. With "an atom is a miniature solar system," the "miniature solar system" would be the "vehicle," the "theme," or the "frame." To illustrate this point further, we could call it the "anchor." In this type of a theory of metaphor, the "anchor" is the literal meaning that enables the metaphor. However, if there is a literal meaning, then not all language is metaphoric. Still, Perelman and Olbrechts-Tyteca (1969) and Richards (1936) proclaimed all language to be metaphoric, and Nietzsche (1990) would agree.

Nietzsche (1990), as I have noted, does not allow the idea of a literal language that is not metaphoric. Indeed, he alludes to science as he diminishes language to vibrations upon a strand of nerve. According to Nietzsche, the source of language is in experience, and his example of the differences and similarities of different types of leaves is one drawn from the natural world. For Nietzsche, what others would call a literal language is merely a matter of "illusions which we have forgotten are illusions" (1990, p. 891). Nietzsche's scientist is a laborer in the semantic hive, using metaphors and hyphenated constructions to generate new meaning. What separates us from animals, according to Nietzsche, is that we have forgotten that our language is a construction and that our belief is based upon lies we have told ourselves about a literal reality that is much easier to believe than a metaphorical one. Though Nietzsche sees such portended knowledge as "the pride connected with knowing and sensing . . . like a blinding fog over the eyes and senses of men, thus deceiving them over the value of existence" (p. 889), the answer to this dilemma is to recognize the metaphors we live by, as Lakoff and Johnson (1980) would recommend. When Nietzsche (1989) aphorizes that "language is rhetoric, because it desires to convey only a *doxa* . . . not an *episteme*" (p. 23), he means to divorce his readers from the notion of words having a literal meaning and to support the idea that the source of words is found in experience. Knowledge for Nietzsche, then, is built on experience.

Nietzsche is not completely alone. Though Berggren (1962/1963) has proposed the idea of a "stereoscopic vision" as a way necessary to view metaphor, which suggests seeing the metaphor and simultaneously being aware of a literal truth, he concludes that some ideas are expressible only as metaphors and that something occurs in the conjunction of ideas that cannot be accounted for with the concept of a literal language. Indeed, for Berggren, the danger of a metaphor is the extent to which it can become a myth. If metaphor can become myth (we certainly know that it can, from the myths of primitive culture to the myths of scientific culture), then the ramifications for metaphor's power are tremendous and add importance to this study. But again, Berggren's "stereoscopic vision" suggests there is a literal meaning upon which the metaphorical meaning is based.

For Nietzsche, the image, not the word, is primary. Arbib and Hesse (1986) have taken this idea one step further. Their schema theory is a construction of ideas independent of imagery. Without imagery, we are left with how to build intelligence, and metaphor serves as a bridge between schemas. It is also significant that Arbib and Hesse agree that all language is metaphoric, but their arrival at this conclusion is not fraught with contradiction, as it is for the others. For Arbib and Hesse, language is metaphoric and not dependent upon images but rather generated by ideas.

I have presented how the solar system analogy played out in the work of physicists over a period of about 50 years and offered evidence of the value of metaphor and analogy to Cambridge physics as well as in the Scottish universities, concluding with Bohr's 1913 publication of his trilogy. In that study, I demonstrated how an analogy became useful to a group of scientists. I also argued that Bohr's rejection of the classical model, the Solar System Analogy, caused the use of metaphor and analogy to be somewhat discredited in scientific thought. What has been the effect on metaphor in science today?

To answer this question, I will shift from physics to biology, specifically to the cloning controversy. Embracing another science strengthens the value of metaphor in the sciences. I will proceed by first describing some of the problems associated with the public reception of cloning, in the political arena and in the public's imagination. Then I will examine documents surrounding the cloning of the sheep Dolly in 1995. This cloning is important because it was the first time that a scientist cloned an animal from a cell other than an embryonic one. Cloning from an embryonic cell is little different than *in vitro* fertilization. Instead, Dolly was cloned from an udder cell. My focus will be on observing how words shift from metaphors to dead metaphors, which means they are becoming accepted as language. In addition, I will consider the effect these metaphors have on other scientists and the public to further strengthen the case for the value of metaphor in a current debate.

THE QUESTION OF CLONING

The debate over cloning has been compared to other medical technology innovations, such as the pacemaker, the artificial heart, organ transplants, and blood transfusion, procedures that we now take for granted. According to a recent Gallup Poll (2006), the percentage of those surveyed who feel that cloning is "morally acceptable" has increased from 52% to 61% over the last four years. The United States Congress has recently eased some restrictions on cloning by at least allowing researchers to use rejected embryos from fertility clinics, but there are still restrictions on federal funding for cloning research.

To some degree, ethics are driving science. For example, recent innovations in gathering stem cells include a process that commingled a human embryonic stem cell with an adult skin cell. As a result, "the embryonic stem cell 'reprogrammed' the skin cell's nucleus, causing the skin cell to start behaving like a youthful, embryonic stem cell" (Check, 2005). More recently, German researchers have cultured cells taken directly from the testes of mice and derived stem cells from them. This process could conceivably allow men to have an inexhaustible supply of stem cells that could be tailored to provide organs that are not only used for replacement, but can also be enhanced. After all, who would want a clone of a defective organ, possibly the one that is being replaced to start with? Unfortunately, such an advantage would work best only for men. For women, the most easily accessible stem cells are in the embryo, which must be fertilized for them to be viable.

In February 2005 the United Nations' legal committee recommended banning all types of human cloning. This decision can be traced to December 2001, when the United Nations General Assembly created the Ad Hoc Committee on an International Convention against the Reproductive Cloning of Human Beings, a group that grew from the decision that reproductive cloning is an "attack on the human dignity of the individual" (Soares, 2003). A consensus agreed that human reproductive cloning was morally wrong, but there was disagreement over the extent that cloning research should be allowed to proceed. Some felt that only reproductive cloning should be banned, but others favored banning cloning of embryonic cells as well.

The cloning of the horse Prometea (Galli et al., 2003), the largest mammal to be cloned, increases the likelihood of human reproductive cloning. Costa Rica introduced a proposal to the United Nations for a ban on all human cloning; most other countries backing this ban are developing countries, those with a Catholic majority population, and the United States. Brazil, it is worth noting, supported banning only reproductive cloning. Conversely, Belgium introduced an alternative ban only on reproductive cloning. The other proponents of this alternative plan include England, France, Germany, and Japan. In addition, on November 6, 2003, the Organization of the Islamic Conference moved in the United Nations that the vote on human cloning be postponed for two years to allow time for further study. This motion carried by a vote of 80 to 79, with 15 states abstaining and other members absent. The United States tried to maneuver the vote to as early as 2004.

To understand the resistance to cloning, it is appropriate to consider how cloning is represented to the public in ways other than the scientific. While reproductive cloning would seem to be ethically wrong for a number of different reasons that I will not explore here, what is most attractive about cloning is what is called stem cell cloning, the process of cloning healthy organs for people who may need a transplant. For those who oppose this type of cloning, the cloning controversy is the same as the abortion controversy, because for such a clone to take place, stem cells must be harvested, and the stem cells must be those present in a viable embryo; that is, one that has been fertilized. The stem cells are undifferentiated, which means they could be used to grow any type of organ. Once the stem cells have been harvested, the embryo is destroyed.

Protest from religious groups that oppose abortion is obvious. How is the idea of cloning playing out in popular culture, though? For it is on this ground that popular opinion may be shaped. One example can be found on the soap opera *The Guiding Light,* where the character Reva was cloned. She was an evil clone who had been created upon her death by her husband, who was evidently a part-time mad scientist. Fortunately, fans posting to Internet message boards expressed incredulity that Reva's adult clone was created within a few weeks and reintroduced to the show. One Web site called "Jumping Shark" uses that phrase to indicate when a soap opera has passed beyond reasonable credibility; those posting to "Jumping Shark" found Reva's cloning to be beyond the believable. Of course, people who post to Web boards are also Web users, and they are not the majority in the United States. As a matter of fact, they are a very small minority. *Alias*, a primetime television show, features an, once again, evil clone of the character Francie. The point is that the idea of cloning is somewhat negative. The clone is seen as evil. Why isn't the clone seen as good as, or even better than, the person who is cloned? The clone is hatched quickly and fully grown into adulthood, which is pure fantasy. Indeed, the clone may have become a symbol for discomfort with technology in general, since it seems to share more with Frankenstein's monster than with the scientific process of cloning.

Another way to consider how cloning is presented to the public would be to examine secondary-school texts. Because the Solar System Analogy was prevalent in many secondary-school texts, it would seem reasonable to find a metaphor for cloning in secondary-school biology texts, so this study next turns to them.

However, an examination reveals no metaphor to describe the process of cloning (Biggs et al., 2002; Feldkamp, 2002; Greenburg, 2001; Kaskel, Hummer, & Daniel, 1999; Miller & Levine, 2002; Starr & Taggart, 2001). Miller and Levine's approach is typical:

> A clone is a member of a population of genetically identical cells produced from a single cell. Cloned colonies of bacteria and other microorganisms are very easy to grow, but this is not always true of multicellular organisms, especially animals. For many years, biologists wondered if it might be possible to clone a mammal—to use a single cell from an adult to grow an entirely new individual that is genetically identical to the organism from which the cell was taken. After years of research, many scientists had concluded this was impossible. In 1997, Scottish scientist Ian Wilmut stunned biologists by announcing that he had cloned a sheep. How did he do it? . . . In Wilmut's technique, the nucleus of an egg cell is removed. The cell is fused with a cell taken from another adult. The fused cell begins to divide and the embryo is then placed in the reproductive system of a foster mother, where it develops normally. . . . Cloned cows, pigs, mice, and other mammals have been produced by similar techniques. Cloned animals are not necessarily transgenic, but researchers hope that cloning will enable them to make copies of transgenic animals that produce genetically engineered substances that will have medical or scientific value (2002, p. 333).

The glossary might be another place to find a metaphoric definition of cloning, but it simply reads, "Clone: member of a population of genetically identical organisms produced from a single cell" (Miller & Levine, 2002, p. 1075). The only metaphors are the "foster mother," an interesting personification, and "fused," a technical metaphor, but neither is the central metaphor describing cloning.

Indeed, the descriptions of cloning by scientists, even the work of Wilmut et al. (1997), who cloned Dolly and Ryuzo Yanagimachi (Wakayama et al., 1999), who used Wilmut's method to clone mice, evidence more metaphor than these textbook examples, and herein lies the paradox: as science pushes further into the unknown, with effects being observed without the cause being understood, language to describe these effects may be more metaphoric. Yet that which is presented to students may be less metaphoric. Where are students to learn of the value of metaphors, and how to use them, other than in the technical communication classroom? Just as technical communication scholars are better prepared to teach writing skills in general to their students, so are they better equipped to teach students about the rhetorical nuances of metaphor and analogy.

Such considerations are interesting in terms of how cloning is represented to the general public, but the question remains as to how scientists are communicating with the public and, to some extent, with each other.

It seems reasonable to expect that such a revolutionary event as cloning would inspire a metaphor to describe it. This metaphor should provide a direction for research and be apparent in other publications. How do other publications treat the metaphor? Do they simply quote it, or do they augment it? If an analogy is invoked, are there other uses of metaphoric language that play into the analogy or spring from it? The answers to these questions will describe the usefulness of metaphor for technical communication as information travels between scientists and then to the public.

I begin examining cloning by looking at ten articles: three from *Nature* and *Nature Biotechnology,* four from *Science*, and three from *Time.* These articles were published after the birth of Dolly, the sheep cloned in 1996 from an udder gland. The articles were chosen from these journals because in terms of audience, *Nature* and *Nature Biotechnology* represent scientific writing for scientists, and *Science*, which tends to report more science news and fewer studies, is situated between these first two journals and the general audience of *Time.* The *Nature* group is certainly one of the most prestigious scientific publications. One of the *Nature* articles is the original "Letter to Nature" where Wilmut et al. (1997) first published his report on the cloning of Dolly. The other articles either report the cloning or comment on it. In a sense, these three categories of journals parallel Einstein's three published theories that describe relativity: one for the physicist, one for the scientist, and one for the general audience.

Over 10 years were spent to produce Dolly. Cloning itself, though it seems "the stuff of science fiction" (Mary had a little clone, 1997, p. 5), is nothing new: in 1952 scientists cloned a frog from embryonic cells, and this procedure has since been applied to mammals—scientists have been cloning the embryonic cells of livestock over the past 10 years. Indeed, cloning has a much longer history with plants—any seedless fruit is a clone.

A clone may be created more easily from embryonic cells because they are undifferentiated, which means they have not yet begun the process of dividing and forming the many different cells from which the mammal will be composed. There is very little difference between this type of cloning and *in vitro* fertilization. The advantage of cloning a non-embryonic cell from an adult is that the cloned animal should be more reliable as an exact copy. Unlike the allusions to playing God, producing hordes of duplicates, creating an extra self for spare parts, or realizing the dream of Dr. Frankenstein, the practical application has been currently most attractive to the livestock industry; instead of being weakened or dying out, a prize line can be replicated *ad infinitum.* Zookeepers, too, could benefit from cloning. Rather than worrying through the tricky process of breeding and pregnancy, which in itself could threaten the life of an endangered animal, zookeepers could clone it. However, cloning is still too risky for the application to be very practical. Cloned animals tend to be genetically weaker and to live shorter lives. More research, then, is necessary for cloning to be useful; a good metaphor to describe it could create a better political environment to gain funding and to prevent legislation against it, as well as be epistemologically generative.

Successful cloning occurred because Roslin Institute researchers, led by Ian Wilmut, shifted to a new technique. For a week, they interrupted cellular division by cutting off sustenance to isolated udder cells, which caused these cells to become inactive. Then, Wilmut implemented nuclear transfer, a standard cloning technique. To perform this process, he removed the nucleus of a cell without disturbing the cytoplasm around it. Next, he positioned another excised nucleus from a different cell, the one to be cloned, next to the first cell whose nucleus had been extracted. Finally, an electrical charge encouraged the cytoplasm to accept the new nucleus, which included the requisite DNA to complete the clone. In this case, the cell that was the basis for the clone regressed to the embryonic state, which signified a successful cloning. After the cell began dividing, it was secured in the uterus of another ewe for eventual live birth.

Dolly's survival was significant because, although scientists have long been able to clone from embryonic cells, this cloning proved that it is possible to clone a cell from a mature mammal; and udder cells, not embryonic cells, were the ones cloned. Dolly's birth, and perhaps more importantly, her survival, promises, unless researchers strike a dead end, to be one of the most important scientific benchmarks of the twentieth century and one that, to draw an analogy with the way Charles Darwin's *Origin of the Species* has influenced the twentieth century, will probably haunt the twenty-first century. Or perhaps a parallel might be more readily drawn with abortion since both are technical procedures, one for creating life and one for destroying it. Or are they? Again, with cloning, the debate over what constitutes life continues.

RECOGNITION OF THE DOMINANT/EMERGENT METAPHORS

Next, the literature must be examined for metaphoric usage. In these articles, the authors used quite a few tropes and figures, which can be classified as

- Metaphor
- Simile
- Hyperbole
- Personification
- Irony
- Cliché

- Pun
- Antithesis
- Metonymy
- Anthimera
- Oxymoron
- Rhetorical question
- Analogy

The examination of metaphor in this analysis is extended to figurative language, a more widely embracing term and one that is often used interchangeably with "metaphor" in this field of study. The various tropes of figurative language serve as the unifying feature of this analysis. The metaphors are further divided and classified as dead, natural, or technical metaphors. It is useful to examine a variety of tropes since any of these can become a dead metaphor, and it is the shift to a dead metaphor that demonstrates the way language evolves.

Dead Metaphors
Gene expression (Wilmut et al., 1997, p. 811)
Colony (Wilmut et al., 1997, p. 812)
Embryonic pattern of expression (Mary had a little clone, 1997, p. 5)
Chimeric embryos (MacQuitty, 1997)
Genetic makeup (Pennisi & Williams, 1997, p. 1415)
Messenger RNA (Pennisi & Williams, 1997, p. 1415)
Cell line (MacQuitty, 1997)
Chromosomal condensation (MacQuitty, 1997)
Compatible cell (MacQuitty, 1997)
Stem cell (MacQuitty, 1997)

An important issue to consider is why a metaphor might be considered dead. Referencing a word in a discipline-specific dictionary is a way to determine the extent that a word or phrase is accepted as a standard term for an object or procedure. To determine the terms listed as dead metaphors for the Dolly study, the *Dictionary of Cell and Molecular Biology* (Lackie & Dow, 1995) and *Henderson's Dictionary of Biological Terms* (Lawrence, 1995) were consulted when those terms or similar ones were found. For example, while "gene expression" was not an entry, "expression cloning" was cited, which was judged close enough to mean that "gene expression" is a dead metaphor since "expression" is now an entry in both biologically-oriented lexicons.

Some metaphors are found in Wilmut's (1997) initial article, but about half are what Johnson-Sheehan refers to as dead metaphors, ideas that "are only definable in terms of metaphor" (Johnson-Sheehan, 1998, p. 169). As examples, Johnson-Sheehan suggests terms such as "force," "energy," "cell," and "space"

(p. 169). These terms once metaphorically described the concepts they now name. In the case of "cell," the word names an object that it now represents. In *Webster's New World Dictionary* (1988), for example, the sixth definition for the word "cell" refers to it as "*Biol.* a very small, complex unit of protoplasm, usually with a nucleus, cytoplasm, and an enclosing membrane: all plants and animals are made up of one or more cells that usually combine to form various tissues." However, the first citation for "cell" defines it as "a small convent or monastery attached to a larger one," and the second definition defines it as "a hermit's hut." For this dictionary, the order of these entries indicates the historical development of the word. How would most people choose if they were asked to consider these three definitions of "cell" and to pick the best one? Dead metaphors other than "cell" are utilized in the initial Wilmut article and in the other articles published after it, especially those that built upon Wilmut's research with other cloning studies.

It is also necessary to differentiate between technical and natural metaphors. Natural metaphors are the type with which most readers are acquainted. These tropes illustrate the unknown entity with the known by associating the unknown with those occurring in the natural world. Some of the natural metaphors are traditional ones, such as the "social and philosophical temblors" (Kluger, 1997, p. 69) and the "ethical shock waves" created by Dolly's birth (Wright, 1997). It is interesting to note that these two compare cloning to cataclysmic, yet natural, forces. Other natural metaphors border on personification since they draw imagery from human relations between individuals and social groups. Such social metaphors include "donor and recipient cells" (Nash, 1997, p. 65; Stewart, 1997, p. 771; Wilmut et al., 1997, pp. 811-812), "regimes [of cells]" (Nash, 1997, p. 65; Wilmut et al., 1997, pp. 811-812), and Wilmut as "Dolly's father" (Marshall, 1997, p. 17). Others such as the "biological barrier" (Nash, 1997, p. 62), are more abstract. One writer felt more at home with a general literary allusion as he noted how "the fiction of this decade becomes the technology of another" (Kluger, 1997, p. 72), one that might seem more naturally to belong with technical metaphor but is included with the natural metaphor because of the traditional literary nature of such an allusion. Though many natural metaphors are present, they are outweighed by technical metaphor.

Technical metaphors explain the unknown in terms of the known by comparing the unknown with figures occurring in the technical world's objects and processes that are the byproducts of science and technology. Variations of one technical metaphor, "programming," which is also expressed as "deprogramming," and "reprogramming," are the most frequently used technical metaphors in these articles (MacQuitty, 1997; Pennisi & Williams, 1997, p. 1416; Nash, 1997, p. 65; Wilmut et al., 1997, p. 812). Borrowed from the computer industry, these terms refer to how the DNA adjusts to being placed in a new embryo. "Remodeling" describes what happens to DNA "in the first cell division," and "packaging proteins" are what the DNA leaves behind in the cytoplasm of the cell it is being cloned from. "Targeting frequency" refers to problems with genes in different types of cells (Stewart, 1997, p. 771). Interestingly enough, "xerox" has made the transition from its original form as a trade name to a verb and an adjective and was expressed in a *Time* article as creating a Xerox copy of someone (Kluger, 1997).

There were a couple of uses of hyperbole that can be thought of as literal and figurative. First, though, it is interesting to note that it is used at all. Of hyperbole, the *Handbook of Technical Writing* comments that, "Used cautiously, hyperbole can magnify an idea without distorting it; therefore, it plays a large role in advertising," but "because technical writing needs to be as accurate and precise as possible, always avoid hyperbole" (Alred, Brusaw, & Oliu, 2000, p. 218).

The first two instances of hyperbole are not terribly strong. According to quoted University of Chicago ethicist Leon Kass, "The NBAC should act . . . as if the future of humanity may lie in the balance" (Marshall, 1997). Though the author of this article did not compose this hyperbolic expression, he did choose it as a quotation to illustrate the debate over cloning.

Another instance of hyperbole is, "In calendar years, seven years from now is a good way off; in scientific terms, it's tomorrow afternoon" (Kluger, 1997, p. 72), which was from *Time*. The *Time* magazine writers used only one instance of hyperbole, and that was, ironically, in relation to time. Perhaps the fantastic nature of cloning struck them as beyond hyperbole. For scientists, the cloning of Dolly is only another step in a series, not a transfer of the fantastic to the natural world but an innovative advancement building upon previous knowledge. Interestingly enough, the authors of the other hyperbole and the metahyperbole were published in *Science* and *Nature Biotechnology*, rather than in *Time,* which is for more general readers and where tropes in general are more likely to be expected to explain science. Both of those usages are also somewhat clichéd, which is more problematic in the sense of becoming a type of dead metaphor that is less useful than the others cited in this study, and even less useful than, say, "chair leg," for example, but they are discussed here as a contrast with metahyperbole.

There was also an instance of what I refer to as metahyperbole, because it constitutes recognition of hyperbole: "if Dolly's announcement brings an influx of money and talent into these aspects of cellular and developmental biology, then all this hyperbole will have been worthwhile" (MacQuitty, 1997). While such recognition of hyperbole is remarkable, so is recognition of metaphor itself as a rhetorical strategy within the scientific community. Scholars of scientific rhetoric such as A. G. Gross (1990) often encounter resistance over even such a basic idea as persuasion as a rhetorical strategy, much less the often-maligned and mistrusted idea of metaphor. This recognition of hyperbole identifies it as a tool of persuasion, quite different from the way metaphors may be begrudgingly seen as useful for instruction or decoration. Of course, an element of persuasion is necessary for effective instruction, a point upon which Socrates and Gorgias agreed (Plato, 1990).

As a trope, personification was so frequently invoked that it is worth examining in the following list:

Donor cells behave (Pennisi & Williams, 1997, p. 1415)

Molecular conversation (Pennisi & Williams, 1997, p. 1416)

The new DNA takes charge (Pennisi & Williams, 1997)

Udder cells are starved (Mary had a little clone, 1997; Nash, 1997, p. 64)

Embryonic tissue forgives (Nash, 1997, p. 64)

Cell cycles are arrested (Nash, 1997, p. 64)

Cells fall into a slumbering state that resembles hibernation (Nash, 1997, p. 64) and are reawakened (p. 65)

The electrical impulse applied to the nucleus to be cloned as the key to getting an egg and donor cell to dance (Nash, 1997, p. 65)

An adult cell has to be coaxed into entering an embryonic state (Nash, 1997, p. 65)

These are examples of the way in which language makes the transition from fresh usages to dead metaphors. Some dead metaphors such as "messenger RNA" began as personified usages, so it would seem reasonable for these personifications to be most present in the writing of scientists, but to the contrary, most of these personifications are garnered from a *Time* magazine writer; only one is drawn from a *Science* writer. None were gathered from the Wilmut et al. (1997) article. The *Time* writer may have been using personification as an audience tool, but why is there such a paucity of personification in the research scientists' prose, especially when the more staid pages of *Nature* and *Nature Biotechnology* are rife with dead metaphors? Where are the "chimeric embryos" (MacQuitty, 1997), "messenger RNA" (Pennisi & Williams, 1997, p. 1415), and "compatible cells" (MacQuitty, 1997)? It would be interesting to know more about how these expressions have come into being. How often, and under what circumstances, do metaphoric terms pass into usage? It would be interesting to know if a Kuhnian (1970) paradigm shift is associated with a proliferation of metaphoric terms. Fahnestock (1999) has called for more research into the development of scientific terms that began as metaphors.

Other tropes were apparent in these articles. A pun is a good example of Berggren's (1962/1963) idea of stereoscopic vision, and in a sense, a pun fits into the condensed figures of speech scheme: if a metaphor is a condensed simile and a simile is a condensed analogy, and analogies themselves can be extended to essay length, then a pun can be thought of as a condensed metaphor. In other words, if the scheme for a metaphor is A:B, for a simile A is like (or as) B, and for an analogy A:B::C:D, then for a pun, the scheme would be A:B, but A is not B, which, like metaphor, requires us to understand the conjunction as well as the disjunction. Of cloned frogs, one writer invoked a pun by quipping that "the best-developing embryos" were "rather ignominiously dying (croaking!) around the tadpole stage" (Stewart, 1997, p. 769). The use of croaking could also be interpreted as onomatopoeia.

Corbett (1990) defines metonymy as the "substitution of some attributive or suggestive word for what is meant" (p. 446). In this case, Kluger advises, "Even the most ardent egalitarians would find it hard to object to an Einstein every 50 years or a Chopin every century. . . ." However, first, science must "get its ethical house in order" (1997, pp. 71-72). Einstein and Chopin represent scientific and musical genius respectively. The biblical allusion to setting one's "house in order" seems a bit clichéd, but the "house" is an "ethical" one, so stylistically it works as a play upon a cliché.

Though it is a figure of speech rather than a trope, there is antithesis worth examining. First, Corbett (1990) defines antithesis as "the juxtaposition of contrasting ideas, often in parallel structure," and one of his examples illustrates antithesis maintained from one sentence to the other. The antithesis in the Dolly article published in *Science* is an allusion to popular culture: "No longer will the name Dolly bring to mind Carol Channing or Barbara Streisand, leading ladies in the musical, 'Hello Dolly,' or even the vivacious country western singer Dolly Parton. Last week, a new Dolly . . . made her debut" (Pennisi & Williams, 1997, p. 1415). Antithesis is usually thought of as contained within a sentence, but similar to a metaphor, it can be extended.

The contrast of ideas in this pop culture allusion again marks the very public interchange with science. It also deals with the frustration of a scientific community eager to begin pursuing research on this new frontier, once it negotiates the general public's fear. By identifying Dolly with Carol Channing, Barbara Streisand, and Dolly Parton, the writer transfers the subject of cloning, the unknown, into the realm of the known, entertainment and popular entertainers. Ironically, the Dolly of *Hello Dolly* is a turn-of-the-century matchmaker. Though certainly lacking in sophistication, Dolly Parton's music and persona exude a crude American wholesomeness. Wilmut himself has identified his Dolly with Dolly Parton by affirming that he named the sheep after her. Perhaps the identification of his cloned sheep with her anticipates resistance to cloning.

Perhaps because the writers of some of these articles may have been taught to disdain and avoid metaphor, there are some clichés apparent, which indicates a lack of competence with metaphor. The *Time* writers tended not to resort to clichés, which points to their professionalism. Clichés in general are ineffective because of their overuse, and this type of plug-in language indicates a lack of practice in the discernment of effective metaphors. One such instance is a description of what happens to the DNA in the cell to be cloned once the nucleus has been transplanted into the new cytoplasm, that it "goes along for the ride" (Pennisi & Williams, 1997, p. 1416). Though this cliché appears in an indirect quotation attributed to Wilmut, the authors of the article chose to use it. What does this mean? Did the authors of the article take this type of cognitive shortcut, or did Wilmut utter it? It would be interesting to know since its use points to another type of dead metaphor. In the same article, developmental cell biologist

Werb is quoted as saying that the cloning of Dolly "is the category of experiment that bends your mind" (as cited in Pennisi & Williams, 1997, p. 1416). This time the cliché is a direct quote. For writing in general, clichés are to be avoided, so it is interesting to find them in articles that report such an innovation as cloning. On the other hand, the Wilmut et al. article (1997) is rife with dead metaphors, and clichés can be thought of as a type of dead metaphor, so perhaps clichés are more acceptable in science writing. However, Wilmut et al. (1997) report a scientific experiment, and there is quite a difference between using terms that are metaphoric and becoming dead metaphors and using clichés. On a similar note, there is quite a difference between clichés and metaphoric terms that become epistemological tools. A cliché does not generate knowledge, and these are ineffective.

Finally, an analogy is almost requisite for a good scientific revolution. Only one appeared in these articles: "A thousand track switches have to click in sequence for the child who starts out toward greatness to wind up there. If one single switch clicks wrong, the high-speed rush toward a Nobel Prize can deadend in a make-shift shack in the Montana woods" (Kluger, 1997, p. 71). This one analogy is not the central metaphor for cloning. Instead, it illustrates the problem with trying to clone genius. Interestingly enough, other meta-phoric elements are embedded in this analogy. It begins with "a thousand track switches," an example of technical metaphor because the railroad imagery is applied not to just the genetic transfer in cloning but also to the cloned genius' environment. Would a clone of Albert Einstein, for example, have to live in an anti-Semitic Europe and work as a patent clerk to be able to contribute to the same extent as the original Einstein? Would the clone of Einstein require the same classical education of his youth? How would any interference affect the outcome?

The train imagery continues in the next sentence, extending the metaphor into an analogy. The use of "Nobel Prize" is an example of metonymy since "Nobel Prize" symbolizes the achievement of genius. "Deadend" functions as anthimera, which is "the substitution of one part of speech for another" (Corbett, 1990, p. 449). In this case, a noun is verbalized and joined as a compound word, but trains do not as often "deadend" as they "derail." A substitution of "derail" for "deadend" would eliminate the anthimera, and it would be more consistent with the train imagery. What follows "deadend" in some ways might justify it because the "makeshift shack in the Montana woods" is an allusion to Ted Kaczynski, the Unabomber. Since this point over "deadend" versus "derail" is arguable, it points to the competence with which the *Time* magazine writers invoke metaphoric language.

Another important point with this analogy is that it is the only one gleaned from these articles, and it is not the defining moment in this scientific revolution, as a good metaphor extended to an analogy should be. Ironically, it questions implementing cloning rather than supporting it.

One other possibility for a central metaphor is Dolly herself. Perhaps Wilmut et al. (1997) were too figuratively tongue-tied to create an effective metaphor, so instead, he presented Dolly, a lamb purportedly named for an American country-western star and associated with a beloved stage character from a more innocent period of American history (at least viewed with the perspective of 100 years). And perhaps the association with American motif was an accurate estimation on Wilmut and his colleagues' part since the United States Senate summoned him to explain his work. However, Dolly could function as an example of synecdoche, which Corbett defines as "a figure of speech in which a part stands for a whole" (1990, p. 445). This definition might fit a little better if it were inverted so that the whole represents the parts. The difference is that in this case, the "parts" are the many different cells that began dividing after the udder cell was cloned.

There are also wider cultural implications for Dolly as a symbol for cloning. In general, sheep bear a positive connotation. According to the biblical account of the birth of Christ, angels announced the event to shepherds, and though there are no direct references to sheep, culturally they have made the transition to countless nativity scenes and church dramatizations. In Judaism, the Israelites enslaved by the Egyptians were instructed to wipe the blood of a lamb on the doorposts of their homes so that the Angel of Death would "pass over" them.

Though Wilmut et al. (1997) shied away from metaphor as a rhetorical strategy, the other authors who described the scientists' work were more lavish in their ornamentation. The number and variety of metaphors within the staid pages of *Nature, Nature Biotechnology,* and *Science* suggest these writers were trying to compensate for the metaphoric deficiency in Wilmut and his colleagues' initial publication, but their use of metaphoric language is probably largely subconscious (though there is some self-conscious usage), and perhaps a reaction to explaining what, indeed, is the true significance of this discovery to the public. Using metaphoric language is perhaps an indication of the writers' desire to interpret the event symbolically. What will cloning ultimately mean to us? They may also be grasping for a metaphor to represent this event.

What is interesting about the use of dead metaphors is that Wilmut et al. (1997) did not use most of them. To be fair, it should be noted that a little less than a third of Wilmut's metaphoric terms were dead, and about two-thirds were ones that can be described as metaphoric. However, such notation refers to five metaphoric expressions, for a total of seven out of approximately 1800 words. Of course, a good metaphor is characterized by the conciseness and precision with which it communicates, not by the number of words, but his article lacks what may be referred to as a central, defining metaphor such as the SSA.

It is interesting that cell biologists, at least from the perspective of this study, seem to feel safer with technical metaphors than with natural metaphors. However, the use of technical metaphors is by no means unique to the description of cloning; technical metaphors in biology have been around at least since the assigning of the Latin names malleus, incus, and stapes, or the hammer, anvil and stirrup, to the three bones in the inner ear.

The *Time* magazine writers tended to use more sophisticated figures of speech, such as metonymy, anthimera, oxymoron, the rhetorical question, and analogy. Such usage may point, again, to the professional writers' competence with metaphor. Not all writers for *Science* are research scientists, but those who write for *Time* are journalists and are certainly in the habit of writing for a general audience. Still, no central metaphor emerged from this study.

THE EFFECT UPON
THE SCIENTIFIC COMMUNITY

With the tropes and figures in these articles identified, I next consider how the scientists reacted to Dolly. The scientific community was quick to test Wilmut and his colleagues' (1997) research. Five months later, University of Hawaii scientists led by R. Yanagimachi (Wakayama et al., 1999), cloned more than 50 mice with Wilmut and his colleagues' method, which Yanagimachi altered by using a needle to perform the transfer of the nucleus. Such a change in the process allowed Yanagimachi to transfer the nucleus without including so much of the cytoplasm from the cell to be cloned, which he believes will prevent further problems from developing after birth. By using mice, Yanagimachi could observe his clones mature and bear normal offspring more quickly. Similar to the Wilmut et al.'s (1997) article, the Yanagimachi study was published in the "Letters to Nature" section. As far as metaphor is concerned, such usage in Yanagimachi's article is somewhat different from Wilmut and his colleagues' in that he uses personification. A direct comparison is shown in Table 2.

First, it is notable that Yanagimachi's (Wakayama et al., 1999) article uses some different metaphoric terms than Wilmut and his colleagues' (1997). This difference can perhaps be accounted for by the contrast in techniques. The use of personification in the Yanagimachi article points to a shift in metaphors in terms of the way cloning is presented to the public, but nothing in this usage suggests conscious usage or exploitation of the metaphor. In both studies, the programming metaphor is present, but Yanagimachi makes less use of it, possibly because his study was based to some degree on Wilmut and his colleagues' work.

Yanagimachi (Wakayama et al., 1999) further altered Wilmut and his colleagues' (1997) approach to cloning by using a needle to inject the nucleus into the egg, rather than using an electrical charge, as Wilmut et al. did, or using a virus or a chemical as other cloning researchers have done, to cause the egg to accept the nucleus. Though Yanagimachi's approach yielded a valuable new approach to cloning, one that became the subject of multiple lawsuits over rights to the process, it yielded no central metaphor to define cloning. Instead, it created a number of metaphoric uses, some in combination with other words in the cloning vernacular, such as "chromatin" in "chromatin repair" and

Table 2. A Comparison of Wilmut and Yanagimachi's Metaphors

Wilmut's metaphors	Yanagimachi's metaphors
Dead Metaphors	**Dead Metaphors**
Colony (p. 812)	Chromosome condensation (p. 370)
Gene expression (p. 811)	Chromosome spindle complex (p. 373)
	Cumulus cells (p. 373)
	Mural granulosa cells (p. 373)
	Polar body (p. 370)
Natural Metaphors	**Natural Metaphors**
Donor and recipient cells	Cumulus-derived chromosomes (p. 370)
(pp. 811-812)	Donor nuclei (p. 369)
Regimes [of cells] (pp. 811-812)	
Technical Metaphors	**Technical Metaphors**
Deprogramming (p. 812)	Electrofusion (p. 369)
Programming (p. 812)	Molecular mechanisms (p. 373)
Reprogramming (p. 812)	Reprogramming (p. 373)
	Personification
	Foster mother (p. 371)
	Trauma (p. 373)

"electrofusion," a coined word. The metaphors for both Yanagimachi and Wilmut are more related to the creation of the process by which new science is produced. Though they can be separated into categories of natural, technical, and scientific, these metaphors identify the process of discovery—the *techne* of epistemology—which is interesting in itself as it relates to how scientific knowledge is created. However, still no central identifying metaphor for cloning emerged.

Another significant cloning that occurred more recently is the 2003 cloning of the colt Prometea (Galli et al., 2003), which is important for its implications for human reproductive cloning since Promotea is the largest mammal to be cloned. Because Dolly was cloned in 1997 and Yanagimachi's mice shortly thereafter, I next turn to the Prometea cloning to observe what changes, if any, took place in the use of metaphor.

For the sake of continuity, articles from most of the same journals were used for the Prometea study. Four articles were examined, and the original study of Dolly was replicated as much as possible. It was not possible to examine the same number of articles from *Nature* or from *Science* because there simply were not as many articles on Prometea in each journal. In fact, two articles on Prometea were published in *Nature*, the Galli et al. (2003) "Brief Communication" (the article written by the scientists who performed the cloning), and a news article, but no article on Prometea was published in *Science*. There were no articles published in *Nature Biotechnology* or *Time*, nor were there articles published in other general newsmagazines, so an article from the *Washington Post* was substituted for the *Time* articles. The *Washington Post* article was chosen based on the newspaper's reputation and the length of the article. The *New York Times* also mentioned Prometea, but only briefly, in an article of fewer than 100 words; the *Washington Post* article was close to 1000 words. The reason for a longer article in the *Washington Post* can be attributed to the political interest in cloning. To increase the number of articles, one from *Science News* is considered.

The sections where the Wilmut and Galli articles were published in *Nature* differ somewhat in terms of status. The Wilmut article was first published as a "Letter to *Nature*." The Galli article was published as a "Brief Communication." The length of a "Letter to Nature" is typically around 1500 words, while a "Brief Communication" is around 500 words. According to the "*Nature* Guide to Authors," "Letters to *Nature* are short reports of original research focused on an outstanding finding whose importance means that it will be of interest to scientists in other fields" (Nature.com, 2003). Though also peer-reviewed, a "Brief Communication" is considered "less formal than Articles and Letters, [and] aimed at the broadest possible readership." A "Brief Communication" should not use more than one visual aid, and the word count should be around 500 words unless there is no visual aid. In that case, a "Brief Communication" may be 700 words. Wilmut's "Letter To Nature" is around 1800 words, and Galli and colleague's "Brief Communication" is around 500 words. A photograph of the cloned colt and her mother accompanies the Galli et al. the article. The Wilmut et al. article features two photographs, a "photomicrograph of donor cell populations," and a photograph of Dolly and the "recipient ewe" (her surrogate mother), as well as two tables. As for the authors of the articles on horse cloning, the only scientists appear to be Galli and his co-authors.

The metaphoric nature of a word was determined by again consulting a discipline-specific dictionary. For the Dolly study, the *Dictionary of Cell and Molecular Biology* and *Henderson's Dictionary of Biological Terms*, both published in 1995, were consulted. For the Prometea study, I consulted *A Dictionary of Genetics*, 6th edition, which was published in 2002 (King & Stansfield). If the terms such as "programming" or "cell remodeling" were not present, then they were considered metaphoric because of their emerging status as a term.

The cloning of Prometea utilized somatic cell cloning, the same technique used to clone Dolly, and also mice, cattle, goats, rabbits, cats, and pigs since then. The difference this time was that not only was a colt produced, a goal that had previously eluded scientists, but Galli et al. (2003) cloned Prometea from a skin cell of the mother in whose womb the colt was nurtured. Being able to do so strengthens the case for the possibility of a woman being able to clone herself.

First, Galli et al. (2003) procured horse ova from a slaughterhouse. Before the nuclei were inserted into the ova, they were emptied of DNA. The nuclei from the mares' skin cells were inserted into the ova and allowed to mature for eight days. Then, they were inserted naturally into the surrogate mothers. Over 800 such ova were prepared, and 22 matured into embryos that were placed into a surrogate mother's uterus. Of those 22 embryos, four were realized as pregnancies, but only one colt was born. Previous clones such as Dolly used DNA from another sheep but were nurtured in the womb of a third, unrelated sheep. Because the colt, who was named Prometea for Prometheus of classical myth, was cloned from the skin cell of the same mare into whose womb she was placed, she was an exact twin to her mother.

This cloning is significant because it is the first time that a clone was brought to fruition from the womb of a female from whom the DNA material had been taken for the clone. "Basically, she foaled herself," noted Galli (as cited in Weiss, 2003). This successful clone upsets a premise of reproduction: that the embryo's DNA must be somewhat different to prevent the host from spontaneously aborting it. The male's DNA has been thought in the past to provide the difference that prevents the spontaneous abortion. Even more important, this cloning now suggests that not only is cloning a human being more likely, but that a woman may be cloned from herself is also possible.

Being able to clone a horse would allow owners of prized lines to prevent them from dying out. Two horse-racing groups have reacted somewhat negatively, however. The Jockey Club, which boasts a 109-year history and serves as the administrative conduit for North American thoroughbred registration, has recently excluded cloned horses from registration or racing, despite the fact that Vice President for Corporate Communications Bob Curran .insists that "If you had ten clones of Secretariat [the 1973 Triple Crown winner], they wouldn't all finish at the same time" (as cited in Weiss, 2003). Many regard Secretariat as the greatest racehorse ever. Cloned horses could participate in other events, such as jumping and harness racing, and cloning horses would be useful for maintaining the lines of endangered wild horses.

Galli himself has noted that the clone of a championship horse may not preserve a great racehorse's ability. The horse's trainer may play a role as well, an observation made in the Dolly study about cloning Albert Einstein. One of the writers on Dolly used metonymy to refer to the possibility of an Einstein or Chopin every generation.

The types of metaphors invoked in the Promotea articles were fewer than the ones used for Dolly:

Dolly Study
- metaphor
- similie
- hyperbole
- personification
- irony
- cliché
- pun
- antithesis
- metonymy
- antihimera
- oxymoron
- the rhetorical question
- analoy

Promotea Study
- metaphor
- hyperbole
- personification
- pun
- synedoche
- allusion

Of these, it is surprising that there were no similes (though a paucity of similes was evident in the Dolly study), especially since a metaphor may be thought of as a condensed simile. Likewise, an analogy has been called an extended simile, and though no coherent analogy was observed to describe the cloning of Dolly or to direct research, no analogy was apparent to describe Prometea's cloning.

With the Prometea study, there were some dead metaphors, such as "genetic identity," but one of the articles on Dolly used the term "genetic makeup." Since these are both dead metaphors, each was listed in one of the discipline-specific dictionaries consulted. Essentially, they mean the same thing. "Adult cells" is interesting as a dead metaphor because "cell" itself is a dead metaphor, and adding "adult" does not change that.

The use of "surrogate mother" is even more telling as a dead metaphor. Though it is listed in *A Dictionary of Genetics* (King & Stansfield, 2002), there is very little mention of the corresponding sheep in the Dolly articles. "Surrogate mother" was one of only two metaphors in the textbook (Miller & Levine, 2002) quoted earlier in this chapter. Wilmut refers to the "recipient ewes," a term that describes a group of ewes, rather than a specific one, so it was not listed in the Dolly study as a metaphor. Most of the other Dolly articles do not mention the recipient ewe at all. Nash (1997) the author of one of the *Time* magazine articles on Dolly, refers to her as a part of the process in which the embryo is "implanted in the uterus of another Blackface ewe." In the Yanagimachi article (Wakayama et al., 1999), the surrogate mother is referred to as the "recipient female" and as a "foster mother." It is important to note that this term is accessible as a dead metaphor, yet it was not used in the Dolly studies. With the Yanagimachi (Wakayama et al., 1999) study, there was a shift to "foster mother." Why is that? In a strictly technical sense, Dolly's "recipient ewe" was more a "surrogate mother" than Prometea's. With a human surrogate mother, a sperm from one source and an egg from another source are united in a petri dish before being placed in a woman who is not the source of the egg, and we refer to the result of this process with the metaphor, "test-tube baby." While the surrogate mother provides nourishment and a place for incubation, she does not share DNA with the maturing embryo. Such was the case for Dolly, but not for Prometea. Prometea's surrogate mother was genetically her mother. This metaphor is inaccurate then, but it represents a type of humanizing of cloning. The term "surrogate mother" is also shared by all four of those who wrote about Prometea. Though Pilcher (2003), the author of the *Nature* news article, mentions surrogacy as describing the relationship between Dolly and her recipient ewe, she flatly calls the recipient mare Prometea's mother; and of these authors, Pilcher was the sole woman, except for some of the scientists in Galli's research group.

The two studies did not share any technical metaphors. The Dolly study contained the metaphors most frequently used, those garnered from computer programming: "programming," "deprogramming," and "reprogramming." These were used by different writers, and most importantly, they were all three used by Wilmut et al. (1997). However, they do not appear in any of the Prometea articles. The reason may be that Galli et al. (2003) are not as compelled to explain their process. They begin their article with, "Several animal species including sheep [Dolly], mice, cattle, goats, rabbits, cats, pigs, and more recently, mules have been reproduced by somatic cell cloning, with the offspring being a genetic copy of the animal donor of nuclear material used for transfer into an enucleated oocyte." Therefore, the description of Galli's process is not required to be as detailed. Certainly Galli let his audience know certain aspects of his approach, such as the chemical triggering of the clone, which is different from how Wilmut used electricity, but there is not as much need for detail. What is so remarkable is not the clone itself but the fact that the mare herself was cloned.

Some of the other technical metaphors are worth commenting on as well. "Spontaneous abortion" is metaphoric because the word "abortion" is defined as any interruption of a pregnancy by intentional interference, so a "spontaneous" abortion places a metaphoric spin upon the term because the mother, or any other outside influence, does not intentionally cause the end of the pregnancy. "Mass reproduction" is a metaphor for the type of fear inspired by cloning, of dictators cloning armies of themselves. The idea of "Dolly overturning an axiom" refers specifically to the belief that a mammal could not be cloned from an adult cell, and an axiom can be read as a human-created concept.

Natural Metaphors
Recipient mare (Galli et al., 2003)
Animal/nuclear donor (Galli et al.,2003)
Empty egg (Pilcher, 2003)
Cloned twin (Pilcher, 2003; Weiss, 2003)
Galli quoted as saying the mare foaled herself (Weiss, 2003)
Harvested mature eggs (Travis, 2003, p. 83)

Immature/mature horse eggs (Travis, 2003, pp. 83-84)
Equine gene pool (Weiss, 2003)

With the natural metaphors, Galli et al. (2003) use the term "recipient" coupled with "mare" as a parallel with Wilmut et al.'s (1997) "recipient ewe." However, Galli also used "surrogate mother," but Wilmut did not, even though it would have more accurately described Dolly's "recipient ewe." There are only two instances of shared metaphor between the two studies: the recipient ewe/mare, which was not defined in the first study, and that of the donor metaphor described metaphorically as a noun and as a verb. Ironically, one *Science* writer (Marshall, 1997), who wrote of Dolly, referred to Wilmut as Dolly's father. It is notable that the types of natural metaphors drawn from the Prometea study are more so related to the cloning process than those drawn from the Dolly study. In the Dolly study, I found more metaphors such as "social and philosophical temblors" (Kluger, 1997, p. 69); "ethical shock waves" (Wright, 1997); "donor and recipient cells" (Nash, 1997, p. 65; Stewart, 1997, p. 771; Wilmut et al., 1997, pp. 811-812;); "regimes [of cells]" (Wilmut et al., 1997, pp. 811-812); "genes as destiny" (Wright, 1997, p. 16); and "biological barrier" (Nash, 1997, p. 62). The Galli et al. (2003) article is about half the length of the Wilmut et al. (1997) article, but it does not provide as much detail on the cloning process. Wilmut et al.'s article uses more technical metaphors to describe the cloning process.

One instance of hyperbole in the Promotea articles is remarkable in that out of the eleven Dolly articles, there were only three instances of hyperbole, so it is somewhat remarkable to find one hyperbole in the four Prometea articles. "A race for survival" seems somewhat clichéd, but it is not categorized as such because Weiss (2003), the *Washington Post* writer, is describing the plight of endangered wild horses.

Again, probably because the process is not as detailed, there are no shared personifications. With the Dolly study, cells behaved, starved, and slumbered, and tissue forgave. Prometea's metaphors do share two instances of personification that contribute to the humanizing of cloning. One is the reference to "celebrity cloning" (Pilcher, 2003), which is not quite the same as other types of personifications when cells were assigned human behavior, and it does not describe the cloning process. Instead, the idea of celebrity is in the eye of the beholder and is a social construction, which could point to a type of maturing of the use of personification as a figure since it is more of an abstraction. Despite this maturation, the use of metaphor and analogy is largely unconscious.

The other personification points out that Cremona, Italy, the birthplace of Prometea, was also "where Antonio Stradivari made his prized violins" (Weiss, 2003). Such a comparison could as easily be categorized as an allusion, but it is a personification in the sense of how it humanizes cloning by associating it with the fine arts.

The synecdoche from the Prometea articles is somewhat different because "hard on the hooves" (Travis, 2003, p. 83) represents an act of research. There was no instance of synecdoche in the Dolly article, but there was one of metonymy, where Einstein and Chopin represented genius.

Allusion was not a category of metaphor in the Dolly study, and it is not traditionally considered a trope or a figure, though I mention it as a natural metaphor in the Dolly study. Certainly an allusion can be considered a rhetorical strategy that has quite a bit in common with metaphor because it requires a comparison between the known (traditionally a literary work) and an interpretation of a more immediate situation. In the Dolly study, a number of allusions were noted as part of discussing figurative elements. There was one of the most traditional of allusions, that of the biblical, as science was instructed to "get its ethical house in order." Interestingly enough, the complete allusion is to "Set thine house in order; for thou shalt die and not live" (Second Kings 20:1). Ironically, there was also an allusion in one of the Dolly study articles to Nietzsche's (1968) "uberman" as an "uberorganism," and many more allusions to cultural figures. In the Prometea study, though there is a popular culture reference to the theme song from the *Mr. Ed* television show (from the lead to the *Washington Post* story), the more important allusion is to classical mythology, the feminization of Prometheus into Prometea. The Galli et al. article offers no explanation, but in the *Washington Post* article, Galli is quoted as saying that, "He [Prometheus] was a brave man. . . . He challenged some conventions and ideas" (Weiss, 2003). Prometheus stole fire from the gods and presented it to humanity. Classical sources differ as to whether Prometheus was a man or a god. Such an allusion has two distinct prongs. One is the humanizing aspect consistent with the Stradivarius violin, the celebrity clones, and Prometea's "recipient mare" as a surrogate mother. The other can be interpreted as a jab at those who would limit cloning research. With the potential to extend human life with organs created by cloning, humanity has the opportunity to make a major advance in the extension of life that rivals the contributions of Pasteur, so who, indeed, is playing God, those who would limit cloning research or those who would explore it?

This study has revealed something of a shift in the use of metaphors. Part of this shift may be explained in terms of the difference of the coverage of the two studies and the amount of detail necessary to convey the process. What is most interesting is that while the role of personification has shifted from describing the process, there is more of a tendency to humanize cloning. When there was a greater necessity to explain "somatic cell cloning," there was more personification of the process. Now that the process has been more widely discussed in the scientific literature, there is a tendency to humanize the process' product. Even though the recipient mare was more so Prometea's twin, three of the four authors used the term "mother" in some sense. Clones such as Dolly and Prometea are becoming "celebrity clones," and Prometea shares her birthplace with Stradivarius violins.

On the other hand, technical metaphors (most of them related to computers, such as "programming" and "deprogramming"), popular in the Dolly study, were not apparent in the Prometea study, but "reprogramming" did appear in the Prometea articles. Such a difference can be accounted for by the fact that the cloning process in the Prometea study was less detailed, but there was less attention paid to how the cloning process occurred because it had been previously documented. However, Prometea was cloned from the mare who nurtured her during pregnancy; therefore, there would seem to be some focus on the idea of the "programming" of the cell, especially since Prometea's mother did not spontaneously abort her. It would seem appropriate to include the computer-inspired metaphors since "abort" as a verb has made the transition to computer science, where it refers generally to ending a process.

Another facet worth mentioning is that responsibility for communicating science to the public lies not only with the scientists but with the science journalists as well. A major responsibility lies with the scientists, though, since Galli et al. (2003) modeled the mother metaphor that the other writers then used themselves.

Metaphor is still used too unconsciously in scientific and technical communication. Greater attention to it would allow better communication between science and the public. Greater attention to metaphors could be epistemologically useful to scientists as well. As E. F. Keller has pointed out, "the object of the verb 'to reprogram' is almost always the nucleus and only rarely the genome. Is there a difference? In fact, there is, and the difference is almost certain to be important" (2000, p. 90). The genome is packaged in the chromatin, and it is there that current research is being directed to discover the source of DNA programming, which is intrinsic to understanding how and why a cell accepts or rejects a nucleus to be cloned. After all, one problem with cloning is that so many attempts fail. Prometea was literally one in about 800.

The use of metaphor is still largely subconscious among scientists. This subconscious use is one that we can connect with theory. The Aristotelian influence can be read in two ways. One is that metaphor for scientists is interactive, as it allows them an opportunity to express the otherwise ineffable. Personification is an obvious example. Another is that rejecting conscious use of metaphor is a rejection of classical education and a validation of the empirical and the positivistic that values laboratory demonstration over theory. Many nineteenth-century scientists found the classical education of their day painfully irrelevant, and metaphor was part of the trappings of a traditional rhetorical education. So it follows that the technical communication classroom is the place to teach about metaphors. And if it is used wisely and consciously, metaphor can be useful to scientists, and such awareness of metaphors can prevent research from being misdirected.

Without a central metaphor, cloning deviates somewhat from the way science has proceeded in the past. In no way should such a conclusion be read as discrediting the use of metaphor in contemporary science. Richard Johnson-Sheehan (1997) has noted how a cluster of metaphors may contribute to the central metaphor and the development of the new concept. In all three studies, there are clusters of metaphors, but no central one. While the writers of these articles utilized some metaphorical expression, none of them contributed to a central image. Instead, these metaphoric expressions seemed to work on a more local level, rather than pointing to a coherent unity.

It is important to note that these metaphors shift from their metaphoric state to being dead metaphors. Traditionally, we think of metaphors as epistemologically generative, and these were probably epistemologically generative in their first incarnations. As dead metaphors, they are epistemologically generative for those who are entering the field, and for those scientists who use them to create new science, but in this sense they are words for objects or processes that have become a part of a larger, more inclusive theory. They are Nietzsche's lies that have forgotten they are lies. Though they are generative only inasmuch as they are used as tools to name and create new theory, they do point to language's metaphoric nature.

It is interesting to reflect upon the metaphorical language and its usage gleaned from this analysis. Why did Wilmut et al. (1997) avoid metaphor as a rhetorical strategy? Why is there no clearly discernible metaphor? Perhaps the reason is that although successfully achieving a clone seems remarkable, his work with Dolly is still in progress. What Wilmut has done so far is the equivalent of Alexander Graham Bell's, "Mr. Watson, come here. I want you!" When Bell uttered these words, the telephone was in its infancy and not ready for practical application. When Wilmut's article (1997) was published, Dolly was also in her infancy, literally, but as a project, she required much more testing and observation to verify her as a successful clone. For example, would she be as healthy as other sheep? If she were not, then what was wrong with her and why? Wilmut would certainly have been interested in how long she lives. Would her life be as long as the life of a normal sheep?[1] Other livestock cloned with embryonic cells have tended to be genetically weak.

Another possibility is that Wilmut may not consider himself the father of a scientific revolution. After all, his successful cloning depended upon a deviation of nuclear transfer, the application of an electrical charge to encourage the egg cell to accept the nucleus to be cloned. He did not invent cloning or nuclear transfer. Still, his application of nuclear transfer and the innovations he applied have enabled cloning to take a giant step forward not possible without his work.

Examining the types of metaphors can explain why no central metaphor emerged. Dead metaphor, natural metaphor, technical metaphor, and personification were the

---

[1] Dolly died February 13, 2003, at the age of six. She was put to sleep because of a lung infection common among sheep who are kept indoors. To prevent her from being stolen, she had been kept all her life at a veterinary school. The average life of a Dorset sheep is 11 to 12 years.

most numerous. Figure 1 compares both the types of metaphor and their repetition by other writers for the Dolly articles. By "repetition of other writers," I mean the extent to which these types of metaphors were used in different articles, which points to their use in the scientific community. Since technical metaphors have the greatest repetition, a question might be posed as to why these technical metaphors are not part of the dead metaphor category. The answer is that a criterion for inclusion in the dead metaphor category required status as an entry in either the *Dictionary of Cell and Molecular Biology* (Lackie & Dow, 1995) or *Henderson's Dictionary of Biological Terms* (Lawrence, 1995).

The technical metaphors that were the most popular were "programming," "deprogramming," and "reprogramming" (MacQuitty, 1997; Nash, 1997, p. 65; Pennisi & Williams, 1997, p. 1416; Wilmut et al., 1997, p. 812). MacQuitty did not use "deprogramming" but he did use the other two. These technical metaphors originated in computer science. Scientists are comfortable with computers, and the United States leads the world in computer use. Ipsos-Insight, a marketing research group, estimates the population of online users in the United States for April 2003 at 135,800, a number that represents only 4.6% of the United States population (United States Census Bureau, 2003). True, there are people who use computers and even the Internet at work and who do not go online, but the point is that these types of technical metaphors may not connect with many people and may alienate them instead. On technical metaphor in molecular biology, specifically those technical metaphors represented by

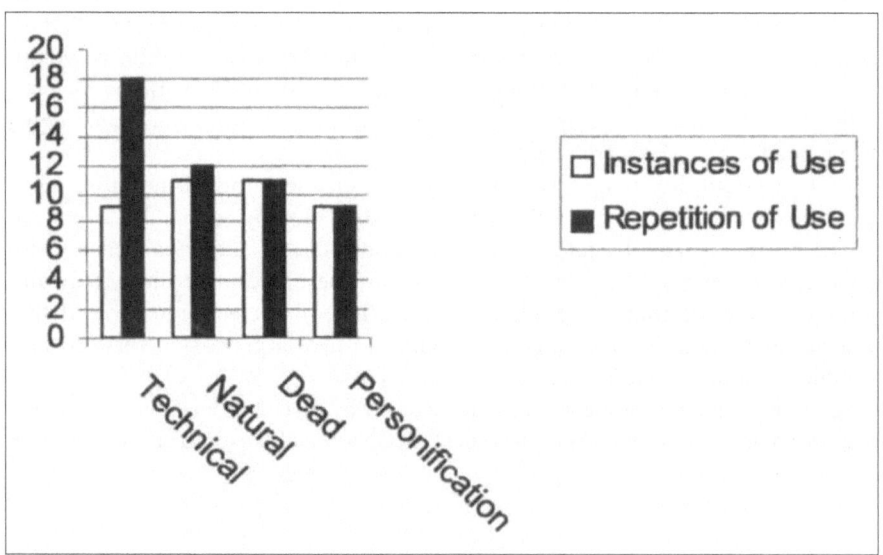

Figure 1. Instances and repetition of use of metaphor.

computers, E. F. Keller has observed that, "without question, computers have provided an invaluable source of metaphors for molecular biology" (2000, p. 81). She has traced this trend back to a 1961 paper by molecular biologists Francois Jacob and Jacques Monod.

As a metaphor, the computer is fertile, and in a sense, it is an extension of the machinery of nature metaphor present since the seventeenth century. As a fertile metaphor, molecular biologists do not know how a cell is reprogrammed after the nucleus to be cloned is placed in it. E. F. Keller (2000) argued that "reprogram" as a metaphor may have misdirected research to the nucleus, rather than the chromatin, where the genome resides. It is here that current research is being directed to discover the source of DNA programming that is intrinsic in understanding how and why a cell accepts or rejects a nucleus to be cloned. After all, one problem with cloning is that so many nuclei must be rejected before one is accepted, so perhaps the reprogramming metaphor may have been misdirecting scientists. Again, metaphor is a powerful tool, but when it is used without discretion and consciousness, it can be misleading. The key is not to avoid metaphor, but to use it self-consciously, which is another reason why it should be taught to fledgling scientists and engineers.

The central metaphor for cloning, then, may be the computer, which is ironic since the computer was first metaphorically described as being like the human brain. However, it is notable that these metaphors do not constitute an analogy comparable to the SSA, but they illustrate how metaphor tends to become embedded in a discipline while an analogy does not do so as readily. The SSA continued to be fertile for Bohr, however, who used it as part of his Nobel Prize acceptance speech in 1922.

Casting metaphor as a *techne* of epistemology is worth discussing. Such an assertion could suggest to some that metaphor should be considered a type of ornamentation. Nothing could be further from the truth. Metaphor in itself is not a philosophy, but it has attracted the attention of such a variety of scholars from disciplines such as rhetoric, philosophy, linguistics, psychology, education, and information technology, and a discipline called "metaphor studies" is emerging. The scholar will continue to turn to metaphor and so will the scientist. It is important to teach the scientist an awareness of metaphor so that it is not something used sloppily or inappropriately. Scientists should consider whether a metaphor has become a myth, for example. On a more local level, the scientist needs to be aware of how the metaphor is working in sentence structure. With a metaphor like "reprogramming," a verb, is it indeed pointing the researcher to the nucleus, rather than to the chromatin? Using metaphor carefully allows it to make its greatest contribution to science.

A central metaphor for cloning would be valuable for communicating with the public. A natural metaphor, rather than a technical one, could smooth the way for the funding of cloning. Another question concerns the extent to which metaphor could direct cloning research. To what extent could the metaphor become the medium? Perhaps cloning could be compared to ways in which the public is accustomed to hearing about genetic manipulation, such as the breeding of pets or the production of flowers. Currently, the question seems to be not how to clone a human being but to understand how the nuclear transfer process works in all its manifestations. Wilmut has not cloned an animal in years; instead, he is working on just this problem, so perhaps the metaphor for cloning might eventually be found in this area. To what extent could plant metaphors be generative? After all, cloning has occurred for much longer with plants than with animals. As a result, it has transparency in our culture. We do not question the ethics of producing seedless fruit, for example. To what extent could the cloning of plants as an analogy be applied to animals? As to whether such an application could be generative to theory, only a scientist can answer that question. Such a metaphor could smooth the path to funding, though.

As a matter of course, the applications of theory should be considered. To what extent is (or should) metaphor be considered from a substitutionist, interactionist, or epistemological perspective? At first glance, the scientific community's attitude seems substitutionist. Metaphor is considered as an aspect of writing that can be simply extracted and ignored, like excessive use of commas, rather than a way in which language and science create themselves. However, scientists are constantly inventing language by using words metaphorically (cumulus cells), casting words together in a phrase (donor nuclei), hyphenating words (cumulus-derived chromosomes), and fusing prefixes and suffixes with words to create new ones (electrofusion) as words shift from a metaphoric to a dead metaphoric usage. Unfortunately, science does not usually credit metaphor with epistemological status. Metaphors are less frequently used consciously than unconsciously. Becoming cognizant of metaphor would allow it to contribute epistemologically.

CONCLUSION

Metaphor did not disappear from science after Bacon's admonitions or even in the twentieth century, as the often-cited work of Watson and Crick's genetic code attests (Gross, 1990; Halloran & Bradford, 1984). This study illustrates that science has not been able to articulate a coherent metaphor for cloning. As a result, the idea of cloning has been left to the popular imagination. If metaphor were taught as a rhetorical strategy in technical communication textbooks, then perhaps these scientists would be more adept at consciously using it.

How did the estate of metaphor fall so low? Many technical communication scholars have traced the source of a distrust of metaphor in the sciences back to Francis Bacon (Baker, 1983; Halloran, 1997; Halloran & Bradford, 1984; Lipson, 1985; Zappen, 1975). However, metaphor was quite alive and well in the nineteenth century, especially in the Scottish universities, and greatly influenced theories of atomic structure. As a matter of fact, Charles Darwin, who attended Edinburgh University for three years while studying to become a medical doctor, used the breeder analogy extensively in *On the Origins of Species*, as J. A. Campbell (1986) has suggested, to make the theory of natural selection more palatable to a nineteenth-century England already well familiar with evolution, perhaps most notably from reading Robert Chambers' *Vestiges of the Natural History of Creation*, as J. A. Secord (2000) has argued.

Metaphor can be useful when scientists attempt to communicate with a general audience, and this general audience is important because it can influence government policy. After all, the United States Senate demanded Wilmut's presence to explain the significance of the cloned sheep Dolly. Traditionally, one of rhetoric's forms has been the legislative, and it is in this arena where a government's policy regarding research has been molded. It is not only where funding is granted or denied, but where limitations can be imposed to ban research altogether. Certainly science has been beset by and survived such attempts at government control, but that does not mean such a path is one science should continue to tread, especially if there are rhetorical (or metaphorical) alternatives.

On the question of a metaphorical versus a literal use of words, this study has demonstrated words shifting from a metaphoric to a dead metaphoric state. Dead metaphors can emanate from a variety of tropes. New knowledge is being created for cloning, and as that knowledge is created, new terms are used to describe it. Those new terms are metaphors until the research community decides to name the object, process, or organism, and then they become dead metaphors. Such an observation is important because here language is being created, which supports the idea that all language is metaphoric.

CHAPTER 6

Implications

This work has supported the idea of teaching metaphor in the technical communication classroom. How such a pedagogical endeavor should be implemented is yet another issue. This chapter first offers some suggestions for teaching metaphor and concludes by considering other implications of this study.

Direction on how to go about teaching metaphor in the technical communication classroom is scant. Jerome Bump (1985) recommended that it be taught and suggested journals as a tool, but he provides little other direction. Though in language studies there is a wealth of material on how to teach students to be aware of and to interpret metaphor, there is little direction on how to teach students how to write metaphors and analogies, other than as it pertains to poetry and other forms of literary writing.

D. M. Catron (1982) provides some structure for teaching analogy to technical writing students. He recommends integrating metaphor into the traditional technical writing assignments, such as the technical description of an ordinary item, the definition of a technical principle, the description of a process or technical instrument, and instructions.

For the technical description of an ordinary item, Catron (1982) suggests that instructors not allow students to use an item from their discipline because they are too likely to know the item too well and rely upon technical jargon. Instead, they should be presented with something they are not quite so familiar with so that they can function more clearly, in a sense, as audience and writer.

For a definition of a technical idea, students choose an idea from their major areas of study. Requiring students to write again for a general audience cuts them off from what Catron refers to as the "perfect audience," their professor in their major and for whom they may have written in the past. He defends this individual as the perfect audience because the professor, "having made the assignment, knows in advance what the answers may be, is thoroughly conversant in the discipline, [and] knows the jargon. Under such circumstances, students scarcely need to communicate at all" (p. 8). In this case, the professor Catron is thinking of seems to be from a discipline other than technical communication. As an exercise to introduce this idea, Catron recommends taking an article from the students' discipline and allowing them to translate it with appropriate metaphors for a lay audience.

Description of a process or technical instrument encourages students to extend the figurative language skills they have practiced to create an analogy to express the idea to a lay audience. Unfortunately, Catron (1982) does not offer much direction on how to guide students to write a good analogy. Instead, he cites one reading in Houp and Pearsall's *Reporting Technical Information* and suggests that students might be allowed to work in groups.

The final assignment suggests that figurative language might be effectively used for writing instructions. At first, Catron seems to recommend that merely the use of figurative language would be helpful to the reader. However, he also suggests as an exercise taking a geometric shape divided into smaller geometric shapes of different colors. The larger shape is then cut up and placed in an envelope. This assignment challenges students to describe how to put the shape back together without being able to rely on jargon or any prior knowledge. As a result, they are encouraged to use metaphor.

Even though Catron focuses on technical communication, his recommendations are problematic. He never suggests that his students are writing analogies, but that they are using figurative language, so semantically, he begins with a problem since there are some differences between figures and analogies. The idea of integrating figurative language into a variety of subjects, beginning with the most basic of technical writing assignments, that of definition, is a good idea. However, a definition is more likely to be a metaphor, and Catron falls short when he faces the idea of implementing longer, more complex pieces such as the analogy, which leads to questions concerning the extent to which his suggestions would be effective. He also does not offer any instances of specific classes that serve as a case study. Instead, his suggestions seem based on classroom experience in general. While such representation of experience certainly has validity, a case study would be more credible to report on specific applications and results.

A better example supported by case studies can be found in B. Sunstein and P. M. Anderson (1989), who have taught the analogy as a research paper to first-year students, but their approach is more focused on scientific writing. After reading science-oriented essays with figurative elements, students write in their journals about the essays. Anderson and Sunstein point out that doing so requires the students to engage epistemologically with the topic, to be the creators of knowledge in this fashion. As they read and write in their journals, the students are especially directed to noting any analogies, figures, or tropes.

Students are encouraged to study a scientific or technical field. They begin with the jargon. They also interview people who know about the topic, and they explore those who are not familiar with it as their targeted audience. They must begin to decide how this discipline is structured in terms of its major and minor fields. Then the students must decide what to compare this scientific discipline to. Another important part of the process involves the shift of the writer to reader. Such a shift demands identification with the audience.

Audience is also important for R. C. Wess (1982), as is the idea of choosing analogy to address a rhetorical issue. In general, he teaches writing by providing his students with examples he has written himself. To teach his students to write an analogy, he drew upon a guest editorial he wrote for a newspaper in a rural area. His editorial sought to explain the importance of writing skills by comparing them to farming.

Emphasizing the prewriting process, he provides a copy of his prewriting strategies that led to his approach to the topic. Such a description is interesting and valuable because at that point in his writing process, Wess had not decided to write an analogy. Hence, the analogy assignment is presented as a choice.

Teaching students to write the analogy seems to require more structure than teaching them to write other types of assignments. Wess provides that structure by triangulating the process. He examines his students' essays, surveys them with questionnaires on the analogy composing process, and assigns a later essay on the process. He also provides two series of heuristic questions, one focusing on the concept of the communication triangle and the other on problem-solving techniques related to the analogy. The communication questions triangulate the writer's knowledge of the topic, awareness of audience, the writer's relationship to audience, and the analogy itself. The problem-solving questions focus on picking the best analogy, one that will be effective for the audience and the writer (p. 9).

Wess's (1982) students were first-year writing students who were taking their first college writing course, so the relevance of his results with regard to technical writing are limited. One architectural engineering student compared baking a cake and constructing a building. His audience was people with no knowledge of this type of construction. To shift frames of reference, another student who worked in the financing office of a department store analogized her job with a circus. As a result, she reported that she enjoyed her job more. The students also seemed to benefit from the revision process required to engage in an analogy.

From the oral presentation perspective, Laura Gurak's (2000) *Oral Presentations for Technical Communication* suggests how students can learn more about metaphor. These suggestions include consciously considering relevant analogies to explain an unfamiliar concept to children or adults. Another directs students to "perform a search on the Web for a scientific or technical topic that interests you. Locate ten to twelve sites related to the topic and look at these for uses of analogies." This exercise is interesting since "the Web" is a metaphor itself. Another assignment requires students to "attend a lecture on a scientific and technical topic" and identify the metaphors. A related, more detailed assignment directs students to interview professionals in their field on the role of analogy in their work. Finally, international communication problems with analogies are addressed by requiring students to interview international students for their perspective on metaphor (pp. 189-190).

While Gurak's exercises are valuable because they work toward creating consciousness of metaphor, students are not given much direction on how to create metaphors and analogies or even how to identify them. Analogies themselves are much less prevalent than metaphors. How do international students react to molecular biology's personifications in English, for example? Clearly, students would require more direction on how metaphors and analogies are created

Some of Wess' (1982) and Sunstein and Anderson's (1989) techniques might prove fruitful. Perhaps Wess' heuristic questions would be valuable. From Anderson and Sunstein's suggestions, the models and the warnings about how an analogy can careen out of control are noteworthy cautions.

AN APPROACH BASED ON THIS STUDY

This work has revealed a wealth of material that would benefit both the student and the instructor. For the introduction to technical communication class, the primary beneficiary, in the sense of who would profit most fully from this study, would be the instructor, who would benefit from the background as an impetus to teach metaphor and analogy in the technical communication classroom.

Some reading can be drawn for students in introductory courses. The instructor should consider the prospective students' scientific background. If most of them are likely to have had a physics course, then selected readings from Descartes, Newton, and Young can illustrate how metaphor shaped the conception of light. For civil engineering students, Smeaton's Eddystone Lighthouse can be detailed. For students with a background in chemistry or physics, the SSA would be relevant. Many of the readings on the SSA drawn upon for this work were from sources intended for general audiences. In this case, students can be shown how the tenets of Scottish Natural Philosophy included metaphor and how its alienation from science was cultural rather than epistemological.

After students are given some background on metaphor, they can be further introduced to it in conjunction with definition. They can be divided into groups and asked to write some definitions, without being told to use metaphors. Then the definitions can be examined for evidence of metaphor, and the distinctions between the definitions that use metaphors and those that do not can be discussed. This approach works well with students who are from specific disciplines. For example, if a class can be broadly broken into various engineering disciplines, then the students can be asked to think of 10 terms relevant to their discipline that would be valuable for a new student to learn. And, of course, metaphor can also be praised in individual student papers.

The lack of research in this area of technical communication in itself points to possible fruitful research in the teaching of analogy in technical communication.

Now that this work has provided so much background to justify the teaching of metaphor in the technical communication classroom, such teaching should be carried out, and the results could be measured through case studies. One research question could pose whether or not the material in this work has been effective in terms of encouraging students to use metaphor, and another research question could measure the quality of student writing to determine if it improves after students are taught to use metaphor.

OTHER AVENUES FOR RESEARCH

There are other possible applications to the classroom. How and what students are taught about metaphors and analogies is a way of differentiating between the technical communication classroom and the business communication classroom. Students from scientific and technical degree programs will use metaphors epistemologically, whereas business students are more likely to use metaphors for communication. Should the methods of teaching metaphor differ?

Further examination of the role of metaphor and analogy might be studied more carefully on several different levels. First, it would be interesting to compare major scientific works from the nineteenth century to those of the twentieth century to determine the extent of the decline of metaphor, and it would interesting to examine the last 10 years to measure usage. Might the metaphors of the computer industry have had an osmotic effect on science in general? In other words, is metaphor usage continuing to decline or to increase?

Second, the last 50 years have seen a boom in technical communication in terms of the support of the complexity of products produced. What is the state of metaphor in technical communication? The last 10 years might be compared with the last 50 years. To what extent has the use of metaphor in the computer industry influenced technical communication? Another study might survey scientific revolutions of the twentieth century to determine if the lack of a metaphor to describe cloning is a fluke or a trend. Differentiation between scientific and technological metaphor would be informative as well. Regardless, the study of metaphor in technical communication should continue to be fruitful.

Finally, new developments in theory are worth noting, in all their manifestations. As I have noted, "metaphor studies" is a discipline emanating from a number of academic areas. These theoretical constructs include empirical studies of metaphor, with which psychology and linguistics are rife, as well as those from philosophy, sociology, and rhetoric. Since artificial intelligence, to some degree, drives this area of study, metaphor as an area of inquiry in technical communication should continue to be fertile.

Perhaps the lack of a central metaphor in scientific theory is more the result of the extensively corporate, or cooperative, nature of contemporary science. However, the lack of direction in the technical communication classroom may also be a factor, and that problem can be traced to a lack of consistency in rhetorical theory and pedagogical this book has sought to address.

References

Alley, M. (1996). *The craft of science writing* (3rd ed.). New York: Springer.

Alred, G. , Brusaw, C. T., & Oliu, W. E. (2000). *Handbook of technical writing* (6th ed.). Boston: St. Martin's.

Anderson, P. V. (1999). *Technical communication: A reader-centered approach.* New York: Harcourt.

Anonymous. (1990). Rhetorica ad herennium, book iv. In P. Bizzell & B. Herzburg (Eds.), *The rhetorical tradition: Readings from classical times to the present* (pp. 252-292). Boston: Bedford.

Arbib, M. A., & Hesse, M. (1986). *The construction of reality.* New York: Cambridge University Press.

Aristotle. (1952a). Logic. Prior analytics. Aristotle I. (A. J. Jenkinson, Trans.). In R. M. Hutchins (Ed.), *Great books of the Western world* (Vol. 8, pp. 139-223). Chicago: Encyclopedia Britannica.

Aristotle. (1952b). On poetics. Aristotle II. (I. Bywater, Trans.). In R. M. Hutchins (Ed.), *Great books of the Western world* (Vol. 9, pp. 681-699). Chicago: Encyclopedia Britannica.

Aristotle. (1991). *Aristotle on rhetoric: A theory of civic discourse.* (G. A. Kennedy, Trans.). New York: Oxford University Press.

Atomic Energy Commission. Retrieved March 18, 2003 from http://energy.gov/aboutus/history/photos.html.

Atoms. (1998). *World book encyclopedia.* CD-ROM. Macintosh ed. Disc 1. USA: IBM.

Bacon, F. (1952). The advancement of learning. In R. M. Hutchins (Ed.), *Great books of the Western world* (Vol. 30, pp. 1-101). Chicago: Encyclopedia Britannica.

Baker, C. (1983) Francis Bacon and the technology of style. *The Technical Writing Teacher, 10,* 118-123.

Beardsley, M. C. (1962). The metaphorical twist. *Philosophy and Phenomenological Research, 17*(8), 293-307.

Berggren, D. (1962/1963). The use and abuse of metaphor, I & II. *Review of Metaphysics, 16,* 237-238, 450-472.

Berlin, J. A. (1987). *Rhetoric and reality: Writing instruction in American colleges, 1900-1985.* Carbondale and Edwardville, IL: Southern Illinois University.

Biggs, A., Daniel, L., Lederman, N., Ortieb, E., Rillero, P., & Zike, D. (2002). *Life science*. Columbus, OH: Glencoe/McGraw-Hill.

Bizzell, P., & Herzberg, B. (Eds.). (1990). *The rhetorical tradition: Readings from classical time, to the present*. New York: St. Martin's Press.

Black, M. (1962). *Models and metaphors*. Ithaca, NY: Cornell University Press.

Black, M. (1978). Afterthoughts on metaphor, how metaphors work: A reply to Donald Davidson. In S. Sachs (Ed.), *On metaphor* (pp 181-192). Chicago: University of Chicago Press.

Blickle, M. D., & Passe, M. E. (1963). *Readings for technical writers*. New York: Ronald Press.

Bohr, N. (1913). On the constitution of atoms and molecules, part 1; Systems containing only a single nucleus, part 2; Systems containing several nuclei, part 3. *Philosophical Magazine, 26,* 3-25, 476-502, 857-875.

Bohr, N. (1958). *Atomic physics and human knowledge*. New York: Science Editions.

Bohr, N. (1981). *Collected work* (Vols. 1-2). L. Rosenfeld (Ed.). New York: North-Holland Publishing Company.

Boyd, R. (1993). Metaphor and theory change: What is 'metaphor' a metaphor for? In A. Ortony (Ed.), *Metaphor and thought* (2nd ed., pp. 481-532). New York: Cambridge University Press.

Bream, S. (2000). Metaphor stacking and the velveteen rabbit effect. STC Proceedings, 378. Retrieved Jan. 29, 2003, from http://www.stc.org/proceedings/ConfProceed/2000.PDFs/ooo66.PDF.

Britton, W. E. (1965). What is technical writing? *College Composition and Communication, 16,* 113-116.

Burke, K. (1960). *A grammar of motives*. Los Angeles: University of California Press.

Bump, J. (1985). Metaphor, creativity, and technical writing. *College Composition and Communication, 36,* 444-453.

Campbell, J. (2001). Rutherford—A brief biography. Retrieved April 9 2003, from http://rutherford.org.nz/biography.htm#top.

Campbell, J. A. (1986). Scientific revolution and the grammar of culture: The case of Darwin's Origin. *The Quarterly Journal of Speech, 72*(4), 351-376.

Catron, D. M. (1982). The creation of metaphor: A case for figurative language in technical writing classes. *Journal of Advanced Composition, 3*(1-2), 69-78.

Check, E. (2005, June 27). Rebooted cells tackle ethical concerns. News at Nature.com. Retrieved July 14, 2005, from doi:10.1038/news050627-1.

Chisholm, R. M. (1986). Selecting metaphoric terminology for the computer industry. *Journal of Technical Writing and Communication, 16*(3), 195-220.

Corbett, E. P. J. (1990). *Classical rhetoric for the modern student* (3rd ed). New York: Oxford University Press.

Crick, F. H. C., Griffith, J. S., & Orgel, L. E. (1957). Codes without commas. *Proceedings of the National Academy of Sciences of the United States of America, 43,* 416-421.

Day, R. A. (1995). *Scientific English: A guide for scientists and other professionals* (2nd ed.). Phoenix, AZ: Oryx.

Davis, E. A., & Falconer, I. J. (1997). *J. J. Thomson and the discovery of the electron*. Bristol, PA: Taylor and Francis.

de Man, P. (1979). *Allegories of reading: Figural language in Rousseau, Nietzsche, Rilke, and Proust.* New Haven, CT: Yale University Press.

Dobrin, D. (1983). What's technical about technical writing? In P. V. Anderson, R. J. Brockmann, & C. R. Miller (Eds.), *New essays in technical and scientific communication: Research, theory, and practice* (pp. 227-250). Amityville, NY: Baywood.

Doppler Effect. (1998). *World book encyclopedia.* CD-ROM. Macintosh ed. Disc 1. USA: IBM.

Einstein, A. (1961). *Relativity: The special and general theory.* (R. W. Lawson, Trans.) New York: Crown Publishers.

EyeSearch.com. (1998). *Optical illusions.* Retrieved September, 24 2003, from http://www.eyesearch.com/optical.illusions.htm.

Fahnestock, J. (1999). *Rhetorical figures in science.* New York: Oxford University Press.

Feather, N. (1973). *Lord Rutherford.* London: Priory Press.

Feather, Jr., R. M., Snyder, S. L., & Hesser, D. T. (1993). *Earth science.* Westerville, Ohio: Glencoe.

Feldkamp, S. (Ed.). (2002). *Modern biology.* Atlanta: Holt.

Frank, D. V., Little, J. G., Miller S., Pasachoff, J. M., & Wainwright, C. L. (2001). *Physical science.* Glenview, IL: Prentice Hall.

Flesch, R. (1951). *How to test readability.* New York: Harper.

Friedland, A. J., & Folt, C. L. (2000). *Writing successful science proposals.* New Haven, CT: Yale University Press.

Galli, C., Lagutina, I., Crotti, G., Colleoni, S. Turini, P. Ponderato, N., Duchi, R., & Lazzari, G. (2003, August 7). A cloned horse born to its dam twin. *Nature, 424,* 635.

Gallup. (2006). *Gallup's pulse of democracy: Stem cell research.* Retrieved July 6, 2006, from http://poll.gallup.com/.

Gamow, V. H. (1964). *The structure of atoms.* Toronto: Macmillan.

Gentner, D., & Gentner, D. R. (1983). Flowing waters or teeming crowds: Mental models of electricity. In D. Gentner & A. L. Stevens (Eds.), *Mental models* (pp. 99-129). Hillsdale, NJ: Erlbaum.

Geiger, H. (1910). The scattering of the α particles. *Proceedings of the Royal Society, A83,* 492-504.

Geiger, H., & Marsden, E. (1909). On a diffuse reflection of the α-particles. *Proceedings of the Royal Society, 82,* 495-500.

Gerson, S. J., & Gerson, S. M. (2003). *Technical writing: Process and practice* (4th ed.). Upper Saddle River, NJ: Prentice Hall.

Goldman, M. (1983). *The demon in the aether.* Edinburgh, Scotland: Paul Harris Publishing.

Gould, J. R. (1979). Bringing teachers of technical writing and teachers of literature closer together. *Journal of Technical Writing and Communication, 9*(2), 173-183.

Greene, B. (2003). *The elegant universe: Superstrings, hidden dimensions, and the quest for the ultimate theory* (2nd ed.). New York: Vintage.

Greenburg, J. (Ed.). (2001). *BSCS biology: A molecular approach* (8th ed.). Chicago: Everyday Learning Corporation.

Gross, A. G. (1990). *The rhetoric of science.* Cambridge, MA: Harvard University Press.

Gurak, L. J. (2003). Toward consistency in visual information: Standardized icons based on task. *Technical Communication, 50*(4) 492-496.

Gurak, L. J. (2000). *Oral presentations for technical communication.* Needham Heights, MA: Allyn and Bacon.

Halloran, S. M. (1997). The birth of molecular biology. In R. A. Harris (Ed.), *Landmark essays on rhetoric of science: Case studies* (pp. 39-50). Mahwah, NJ: Erlbaum.

Halloran, S. M., & Bradford, A. N. (1984). Figures of speech in the rhetoric of science and technology. In R. J. Connors, L. S. Ede, & A. A. Lunsford (Eds.), *Essays on classical rhetoric and modern discourse* (pp. 179-192). Carbondale, IL: Southern Illinois University Press.

Halloran, S. M., & Whitburn, M. D. (1982). Ciceronian rhetoric and the rise of science: The plain style reconsidered. In J. J. Murphy (Ed.), *The rhetorical tradition and modern writing* (pp. 58-72) New York: Modern Language Association of America.

Harman, P. M. (Ed.). (1985a). *Wranglers and physicists: Studies on Cambridge physics in the nineteenth century.* Dover, NH: Manchester University Press.

Harman, P. M. (1985b). Edinborough philosophy and Cambridge physics: The natural philosophy of James Clerk Maxwell. In P.M. Harman (Ed.), *Wranglers and physicists: Studies on Cambridge physics in the nineteenth century* (pp. 202-224.) Dover, NH: Manchester University Press.

Harmon, J. E. (1986). Perturbations in the scientific literature. *Journal of Technical Writing and Communication, 16*(4), 311-317.

Harmon, J. E. (1994). The uses of metaphor in citation classics from the scientific literature. *Technical Communication Quarterly, 3,* 179-194.

Harris, J. S. (1975). Metaphor in technical writing. *The Technical Writing Teacher, 2*(2), 9-13.

Harris, J. S. (1986). Shape imagery in technical terminology. *Journal of Technical Writing and Communication, 16*(1/2), 55-61.

Harris, J. S. (1993). Poetry and technical writing. *Journal of Technical Writing and Communication, 23,* 313-331.

Hays, R. (1961). What is technical writing? In D. H. Cunningham and H. A. Estrin (Eds.). *The teaching of technical writing* (pp. 3-8). Urbana, IL: NCTE.

Heilbron, J. L. (1985). Bohr's first theories of the atom. In A. P. French and P. J. Kennedy (Eds.), *Niels Bohr: A centenary volume* (pp. 33-49). Cambridge, MA: Harvard University Press.

Hesse, M. (1970). *Models and analogies in science.* Notre Dame, IN: University of Notre Dame Press.

Hewitt, P. G. (1992). *Conceptual physics: The high school physics program* (2nd ed.). New York: Addison-Wesley.

Hobbes, T. (1983). *Leviathan.* Middlesex, England: Penguin.

Hodges, J. C., Horner, W. B., Webb, S. S., & Miler, R. K. (1998). *Harbrace college handbook* (Rev. 13th ed.). New York: Harcourt.

Holden, C. (Ed.). (1997). Mary had a little . . . clone. *Science, 275,* 1271.

Holton, G. (1986). *The advancement of science and its burdens.* New York: Cambridge University Press.

Houp, K. W., Pearsall, T., Tebeaux, E. & Dragga, S. (2002). *Reporting technical communication* (10th ed.). New York: Oxford University Press.

Hume, D. (1964). *A treatise of human nature: Being an attempt to introduce the experimental method of reasoning into moral subjects and dialogues concerning natural religion.* Darmstadt, Germany: Scientia Verlagaler.

Hunt, B. J. (1991). *The Maxwellians.* Ithaca, NY: Cornell University Press.

Hunt, K. W. (1977). Early blooming and late blooming syntactic structures. In C. R. Cooper & L. Odell (Eds.), *Evaluating writing: Describing, measuring, judging* (pp. 91-106). Urbana, IL: NCTE.

Hull, K. N. (1980). Notes from the besieged, or why English teachers should teach technical writing. *College English, 41,* 876-83.

Ipsos-Insight Retrieved May 12, 2003 from http://ipsoinsight.com/tech/publications/fow.cfm.

Johnson-Eilola, J. Selber, S., & Selfe, C. (1999). Interfacing: Multiple visions of computer use in technical communication. In T.C. Kynell & M. G. Moran (Eds.), *Three keys to the past: The history of technical communication* (pp. 197-226). Stamford, CT: Ablex Publishing Company.

Johnson-Sheehan, R. D. (1995). Scientific communication and metaphor: An analysis of Einstein's 1905 special relativity paper. *Journal of Technical Writing and Communication, 25*(1), 71-83.

Johnson-Sheehan, R. D. (1997). The emergence of a root metaphor in modern physics: Max Planck's "quantum" metaphor. *Journal of Technical Writing and Communication, 27*(2), 177-190.

Johnson-Sheehan, R. D. (1998). Metaphor in the rhetoric of scientific discourse. In J. T. Battalio (Ed.), *Essays in the study of scientific discourse: Methods, practice, and pedagogy* (pp. 167-181). Stamford, CT: Ablex Publishing Company.

Johnson-Sheehan, R. D. (1999). Metaphor as hermeneutic. *Rhetoric Society Quarterly, 29*(2), 47-64.

Jolly, W. P. (1974). *Sir Oliver Lodge.* London: Constable.

Jones, D., & Lane, K. (2002). *Technical communication: Strategies for college and the workplace.* New York: Longman Press.

Kaskel, A., Hummer, P. J. Jr., & Daniel, L. (1999). *Biology: An everyday experience* (3rd ed.). New York: Glencoe McGraw-Hill.

Keller, E. F. (2000). *The century of the gene.* Cambridge, MA: Harvard University Press.

King, R. C., & Stansfield, W. D. (2002). *A dictionary of genetics* (6th ed.). New York: Oxford University Press.

Kinneavy, J. L. (1971). *A theory of discourse.* Englewood Cliffs, NJ: Prentice-Hall.

Kluger, J. (1997, March 10). Will we follow the sheep? *Time,* 69-72.

Koestler, A. (1964). *The act of creation.* New York: Macmillan.

Kuhn, T. S. (1957).*The Copernican revolution: Planetary astronomy in the development of Western thought.* Cambridge, MA: Harvard University Press.

Kuhn, T. S. (1970). *The structure of scientific revolutions* (2nd ed.) Chicago: University of Chicago Press.

Kuhn, T. S. (1977). *The essential tension: Selected studies in scientific tradition and change.* Chicago: University of Chicago Press.

Kuhn, T. S. (1993). Metaphor in science. In A. Ortony (Ed.), *Metaphor and thought* (2nd ed., pp. 533-542). New York: Cambridge University Press.

Kynell, T. C. (1999). *Scenarios for technical communication: Critical thinking and writing.* Needham Heights, MA: Allyn and Bacon.

Lackie, J. M., & Dow, J. A. T. (Eds.). (1995). *Dictionary of cell and molecular biology* (3rd ed.). New York: Academic Press.

Lakoff, G., & Johnson, M. (1980). *Metaphors we live by.* Chicago: University of Chicago Press.

Lakoff, G., & Núñez, R. E. (2000). *Where mathematics comes from: How the embodied mind brings mathematics into being.* New York: Basic Books.

Lamb, W. G., Cuevas, M. M., & Lehrman, R. L. (1989). *Physical science.* Chicago: Harcourt, Brace, Jovanovich.

Lannon, J. S. (2003). *Technical communication* (9th ed.). New York: Longman.

Lawrence, E. (Ed.). (1995). *Henderson's dictionary of biological terms* (11th ed.). New York: Wiley and Sons.

Lay, M. M., Wahlstrom, B. J., Selfe, C. L. Selzer, J., & Rude, C. (2000). *Technical communication* (2nd ed.). New York: McGraw-Hill.

Leatherdale, W. H. (1974). *The role of analogy, model, and metaphor in science.* Oxford: North-Holland Publishing Co., Ltd.

LeMay, H. E. B., Robblee, K. M., & Brower, D. C. (2000). *Chemistry: Connections to our changing world.* Upper Saddle River, NJ: Prentice Hall.

Lipson, C. S. (1985). Francis Bacon and plain scientific prose: A reexamination. *Journal of Technical Writing and Communication, 15*(2), 143-155.

Locke, J. (1990). From an essay concerning human understanding. In P. Bizzell & B. Herzberg (Eds.), *The rhetorical tradition: Readings from classical times to the present* (pp. 699-710). Boston: Bedford.

Lodge, O. (1902). On electrons. *Journal of the Institution of Electrical Engineers, 27,* 45-117.

Lodge, O. (1907). *Electrons.* New York: Macmillan.

Lodge, O. (1924). *Atoms and rays: An introduction to modern views on atomic structure and radiation.* London: Ernest Benn Limited.

Lodge, O. (1932a). *Advancing science.* New York: Harcourt.

Lodge, O. (1932b). *Past years: An autobiography.* New York: Scribner's.

Loewenberg, I. (1973). Truth and consequences of metaphor. *Philosophy and Rhetoric, 6*(1), 30-46.

Loewenberg, I. (1975). *Denying the undeniable: Metaphors are not comparisons* (pp. 305-316). Mid America Linguistics Conference, Lawrence, KS.

Lowe, P. (1981). The British Association and the provincial public. In R. MacLeod & P. Collins (Eds.), *The parliament of science: The British Association for the Advancement of Science 1831-1981* (pp. 118-144). Middlesex, England: Science Reviews.

MacLeod, R. (1981). Introduction. On the advancement of science. In R. MacLeod & P. Collins (Eds.), *The parliament of science: The British Association for the Advancement of Science 1831-1981* (pp. 17-42). Middlesex, England: Science Reviews.

MacQuitty, J. J. (1997). The real implications of Dolly. *Nature Biotechnology, 15,* 294.

Markel, M. (1998). *Technical communication: Situations and strategies.* New York: St. Martin's.

Marshall, E. (1997). Mammalian cloning debate heats up. *Science, 275,* 1733.

Matthews, J. R., Bowen, J. H., & Matthews, R. M. (1997). *Successful science writing: A step-by-step guide for the biological and medical sciences.* New York: Cambridge University Press.

Maxwell, J. C. (1986a). Atom. In E. Garber, S. G. Brush, & C. W. F. Everitt (Eds.), *Maxwell on molecules and gases* (pp. 176-215). Cambridge, MA: MIT Press.

Maxwell, J. C. (1986b). Draft of atom article for the Encyclopedia Britannica. In E. Garber, S. G. Brush, & C. W. F. Everitt (Eds.), *Maxwell on molecules and gases* (pp. 170-174). Cambridge, MA: MIT Press.

McLaughlin, C. W., & Thompson, M. (1999). *Physical science.* Peoria, IL: Glencoe.

McMullan, E. (1968). What do physical models tell us? In B. Van Kootselaar & J. F. Staal (Eds.), *Logic, methodology and philosophy of science III* (pp. 385-396). Amsterdam: North-Holland.

McMullan, E. (1976). The fertility of theory as the unit for appraisal in science. *Boston Studies in the Philosophy of Science: Essays in Memory of Imre Lakatos, 39,* 681-718.

Miller, C. R. (1979). A humanistic rationale for technical writing. *College English, 40,* 610-617.

Miller, K., & Levine, J. (2002). *Biology.* Upper Saddle River, NJ: Prentice Hall.

Moore, P. (1996). Instrumental discourse is as humanistic as rhetoric. *Journal of Business and Technical Communication, 10*(1), 100-118.

Moore, R. (1966). *Niels Bohr: The man, his science, and the world they changed.* New York: Knopf.

Moran, M. G. (1985). The history of technical and scientific writing. In M. G. Moran & D. Journet (Eds.), *Research in technical communication: A bibliographic sourcebook* (pp. 25-38). Westport, CT: Greenwood.

Morrell, J., & Thackeray, A. (1981). *Gentlemen of science: Early years of the British Association for the Advancement of Science.* Oxford, England: Clarendon Press.

Murphy, J. T., Hollon, J. M. , & Zitzewitz, P. W. (1986). *Physics: Principles and problems.* Columbus, OH: Merrill.

Myers, R. T., Oldham, K. B., & Tocci, S. (2000). *Chemistry: Visualizing matter.* New York: Holt, Rinehart, and Winston.

Nagaoka, H. (1904). Kinetics of a system of particles illustrating the line and the band spectrum and the phenomena of radioactivity. *Philosophical Magazine, 6*(7), 445-455.

Nash, J. M. (1997, March 10). The age of cloning. *Time,* pp. 62-65.

Nature.com. Instructions for Authors. Retrieved December 16, 2003 from http://www.nature.com/nature/authors/gta/index.html.

Neufeldt, V. (Ed.). (1988). *Webster's new world dictionary* (3rd ed.). New York: Simon and Schuster.

Nietzsche, F. (1968). Thus spoke Zarathustra. In W. Kaufman (Ed. & Trans.), *The portable Nietzsche* (pp. 103-439). New York: Penguin Books.

Nietzsche, F. (1989). The relation of the rhetorical to language. In S. Gilman & C. Blair (Eds. & Trans), *Friedrich Nietzsche on rhetoric and language* (pp. 21-25). New York: Oxford University Press.

Nietzsche, F. (1990). On truth and lies in a nonmoral sense. In P. Bizzell & B. Herzberg (Eds.), *The rhetorical tradition: Readings from classical times to the present* (pp. 885-896). Boston: Bedford.

Olson, R. (1975). *Scottish philosophy and British physics 1750-1880: A study in the foundations of Victorian scientific style.* Princeton, NJ: Princeton University Press.

Peacham, H. (1954). *The garden of eloquence.* Gainesville, FL: Scholars' Facsimiles & Reprints.

Pearson, H. (2005, February 22). To know science is to love it. News@Nature.com. Retrieved March 10, 2005, from http://www.nature.com/index.html.

Pennisi, E., & Williams, N. (1997). Will Dolly send in the clones? *Science, 275,* 1415-1416.

Penrose, A. M., & Katz, S. B. (2004). *Writing in the sciences: Exploring conventions of scientific discourse.* New York: Pearson.

Perelman, C. (1982). *The realm of rhetoric.* Notre Dame, IN: University of Notre Dame Press.

Perelman, C. (1989). Analogy and metaphor in science, poetry, and philosophy. In R. D. Dearin (Ed.), *The new rhetoric of Chaim Perelman: Statement and response* (pp. 79-89). Lanham, MA: University Press of America.

Perelman, C., & Olbrechts-Tyteca, L. (1969). *The new rhetoric.* Notre Dame, IN: University of Notre Dame Press.

Petrie, H. G., & Oshlag, R. S. (1993). Metaphor and learning. In A. Ortony (Ed.), *Metaphor and thought* (2nd ed., pp. 579-609). New York: Cambridge University Press.

Pfieffer, W. S. (2003). *Technical writing: A practical approach* (5th ed.). Upper Saddle River, NJ: Prentice Hall.

Pilcher, H. R. (2003, August 7). First cloned horse born. News at Nature.com. Retrieved Dec. 16, 2003, from http://www.nature.com/news/bysubject/developmentalbiology/0308.html.

Plato. (1990). Gorgias. In P. Bizzell & B. Herzberg (Eds.), *The rhetorical tradition: Readings from classical times to the present* (pp. 61-112). Boston: Bedford.

Popper, K. (1972). Science: Conjectures and refutations. In H. Morick (Ed.), *Challenges to empiricism* (pp. 128-160). Belmont, CA: Wadsworth.

Pullman, B. (1998). *The atom in the history of human thought.* (A. Reisinger, Trans.). New York: Oxford University Press.

Rayleigh, L. (1943). *The life of Sir J. J. Thomson.* Cambridge: University Press.

Richards, I. A. (1936). *The philosophy of rhetoric.* New York: Oxford University Press.

Richards, I. A. (1943). *Basic English.* New York: Norton.

Ricoeur, P. (1975). *The rule of metaphor: Multi-disciplinary studies of the creation of meaning in language.* (R. Czerny with K. McLaughlin & J. Costello, Trans.). Toronto: University of Toronto Press.

Rosenfeld, L. (1981). *Niels Bohr: Biographical sketch. Collected work* (Vol. 1). New York: North-Holland Publishing Company.

Rowland, P. (1990). *Oliver Lodge and the Liverpool Physical Society.* Liverpool, England: Liverpool University Press.

Rutherford, E. (1911). The scattering of α and β particles by matter. *The Philosophical Magazine, 21,* 669-688.

Rutherford, E. (1963). The structure of the atom. In J. Chadwick (Ed.), *The collected papers of Lord Rutherford* (Vol. 2, pp. 445-455). New York: Interscience Publishers.

Rutter, R. (1985). Poetry, imagination, and technical writing. *College English, 47,* 698-712.

Ryle, G. (1951). Thinking and language, part III. *Proceedings of the Aristotelian Society for the Systematic Study of Philosophy, Supplement 25,* 65-82.

Secord, J. A. (2000). *Victorian sensation: The extraordinary publication, reception, and secret authorship of Vestiges of the natural history of creation.* Chicago: University of Chicago Press.

Sharlin, H. I. (1979). *Lord Kelvin: The dynamic Victorian.* University Park, PA: The Pennsylvania State University Press.

Smeaton, J. (1953). Selections from a narrative of the building of the Eddystone Lighthouse. In W. J. Miller & L. E. A. Saidla (Eds.), *Engineers as writers* (pp. 90-100). New York: Van Nostrand.

Soares, C. (2003, October 1). UN clone talks bog down. *The Scientist.* Retrieved Dec. 15, 2003, from http://www.biomedcentral.com/news/20021001/05/.

Sprat, T. (1958). History of the Royal Society. J. I. Cope & H. W. Jones (Eds.), St. Louis: Washington University Press.

Starr, C., & Taggart, R. (2001). *Biology: The unity and diversity of life* (9th ed.). Thomson Learning.

Stewart, B., & Tait, P.G. (1876). *The unseen universe or physical speculations on a future state* (6th ed.). London: Macmillan and Co.

Stewart, C. (1997). An udder way of making lambs. *Nature, 385,* 769-771.

Stewart, D. (1829). Elements of the philosophy of the human mind. In *The works of Dugald Steward in seven volumes* (Vol. 2). Cambridge MA: Hilliard and Brow.

Stolberg, R., & Hill, F. F. (1980). *Physics: Fundamentals and frontiers.* Geneva, IL: Houghton Mifflin.

Stoney, G. J. (1891). On the cause of double lines and of equidistant satellites in the spectra of gases. *The Scientific Transactions of the Royal Dublin Society, 4*(2), 563-608.

Sunstein, B., & Anderson, P. M. (1989). Metaphor, science, and the spectator role: An approach for non-scientists. *Teaching English in the Two-year College, 16,* 9-16.

Thomson, J. J. (1883). *A treatise on the motion of vortex rings.* London: Macmillan.

Thomson, J. J. (1903). The magnetic properties of systems of corpuscles describing circular orbits. *Philosophical Magazine, 6,* 673-693.

Thomson, J. J. (1904). On the structure of the atom: An investigation of the stability and periods of oscillation of a number of corpuscles arranged at equal intervals around the circumference of a circle; With application of the results to the theory of atomic structure. *Philosophical Magazine, 7*(39), 237-265.

Thomson, J. J. (1907). *The corpuscular theory of matter.* New York: Scribner's.

Thomson, J. J. (1997). The structure of the atom. In E. A. Davis & I. J. Falconer (Eds.), *J. J. Thomson and the discovery of the electron* (pp. 217-229). Bristol, PA: Taylor and Francis.

Thomson, W. (Lord Kelvin). (1910). Hydrodynamics. In J. Larmor (Ed.), *Mathematical and physical papers, Vol. 4: Hydrodynamics and general dynamics* (pp. 1-68). Cambridge: University Press.

Travis, J. (2003, August 9). Winning bet: Horse and mule clones cross the finish line. *Science News, 164,* 83-84.

United States Census Bureau. (2003). Retrieved May 12, 2003, from http://www.census.gov/cgi-bin/popclock.

Van Alstyne, J. S., & Trit, M. D. (2002). *Professional and technical writing strategies.* Upper Saddle River, NJ: Prentice Hall.

Wakayama, T., Perry, A. C. F., Zuccotti, M. , Johnson, K. R., & Yanagimachi, R. (1999). Full-term development of mice from enucleated oocytes injected with cumulus cell nuclei. *Nature, 394,* 369-374.

Watson, J. D., & Crick, F. H. C. (1953). Genetical implications of deoxyribonucleic' acid. *Nature, 171,* 964-967.

Weaver, R. (1990). The Phaedrus and the nature of rhetoric. In P. Bizzell & B. Herzburg (Eds.), *The rhetorical tradition: Readings from classical times to the present* (pp. 1044-1054). Boston: Bedford.

Weiss, R. (2003, August 7). First cloned horse created in Italy. *The Washington Post,* p. A1. Retrieved Dec. 16, 2003, from LexisNexis Academic.

Wess, R. C. (1982). *A teacher essay as model for student invention.* Paper presented at the 33rd annual meeting of the conference on College Composition and Communication, San Francisco, CA. (ERIC Document Reproduction Service No. ED 217 478.)

Whitburn, M. D., Davis, M., Higgins, S., Oates, L., & Spurgeon, K. (1999). Landmark essay: The plain style in scientific and technical writing. In T. C. Kynell & M. G. Moran (Eds.), *Three keys to the past: The history of technical communication* (pp. 123-130). Stamford, CT: Ablex Publishing Company.

Wilbraham, A. C., Staley, D. D., Matta, M. S., & Waterman, E. L. (2000). *Chemistry* (5th ed.). Glenview, IL: Prentice Hall.

Wilmut, I., Schnieke, A. E., McWhir, J., Kind, A. J., & Campbell, K. H. S. (1997). Viable offspring derived from fetal and adult mammalian cells. *Nature, 385,* 810-813.

Winkler V. M. (1983). The role of models in technical and scientific writing. In P. V. Anderson, R. J. Brockmann, & C. Miller (Eds.), *New essays in technical and scientific communication: Research, history, and practice* (pp. 111-112). Amityville, NY: Baywood.

Wistrom, C., Phillips, J., & Strozak, V. (2000). *Chemistry: Concepts and applications.* Peoria, Illinois: Glencoe.

Wittgenstein, L. (1958). Philosophical investigations (3rd ed.). (G. E. M. Anscombe, Trans.) New York: Macmillan.

Woolever, K. R. (2002). *Writing for the technical professions* (2nd ed.). New York: Longman.

Wright, R. (1997, March 10). Can souls be xeroxed? *Time,* p. 73.

Yagi, E. (1964). On Nagaoka's Saturnian atomic model (1903). *Japanese Studies in the History of Science, 3,* 31-47.

Yeo, R. (1981). Scientific method and the image of science 1831-1891. In R. MacLeod & P. Collins (Eds.), *The parliament of science: The British Association for the Advancement of Science 1831-1981* (pp. 65-88). Middlesex, England: Science Reviews.

Young, R. E., Becker, A. L., & Pike, H. L. (1970). *Rhetoric, discovery, and change.* New York: Harcourt, Brace, & World.

Zappen, J. P. (1999). Landmark essay: Francis Bacon and the historiography of scientific rhetoric. In T. C. Kynell & M. G. Moran (Eds.), *Three keys to the past: The history of technical communication* (pp. 49-62). Stamford, CT: Ablex.

Zappen, J. P. (1975). Francis Bacon and the rhetoric of science. *College Composition and Communication, 26,* 244-247.

Index